The Cambridge Manuals of Science and
Literature

PRIMITIVE ANIMALS

Ornithorhyncus (Platypus)
paradoxus.

Echidna aculeata.

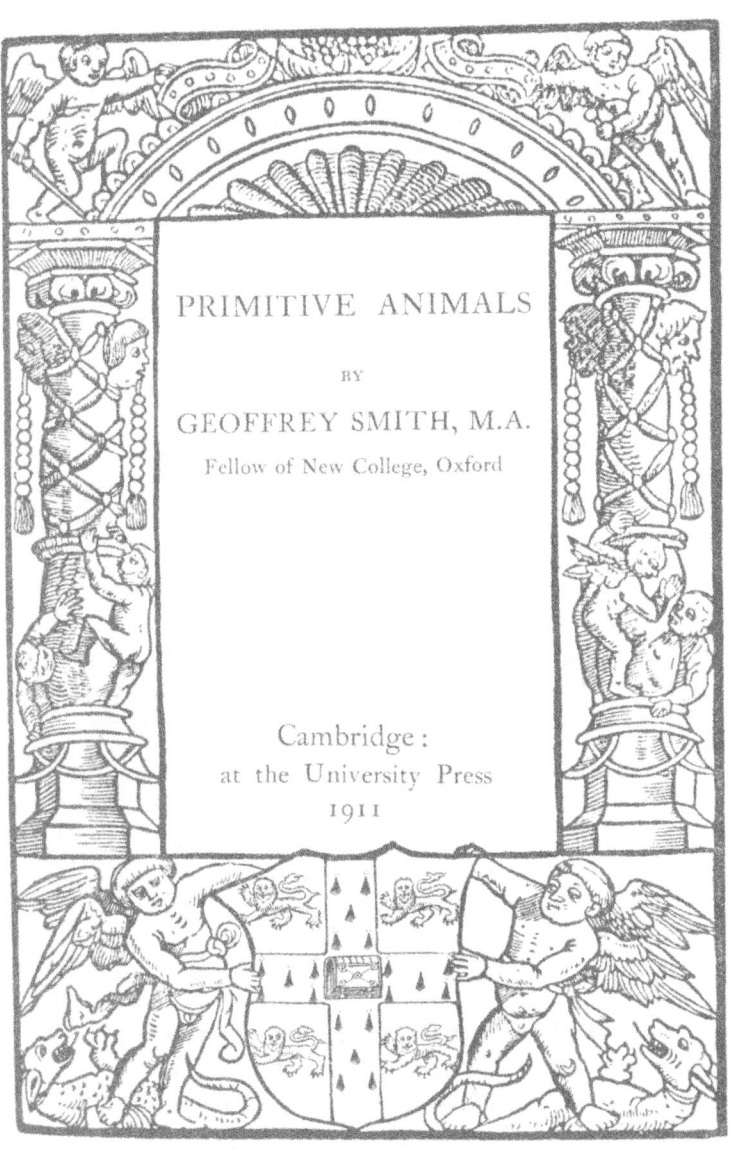

PRIMITIVE ANIMALS

BY

GEOFFREY SMITH, M.A.

Fellow of New College, Oxford

Cambridge:
at the University Press
1911

CAMBRIDGE UNIVERSITY PRESS
Cambridge, New York, Melbourne, Madrid, Cape Town,
Singapore, São Paulo, Delhi, Tokyo, Mexico City

Cambridge University Press
The Edinburgh Building, Cambridge CB2 8RU, UK

Published in the United States of America by Cambridge University Press, New York

www.cambridge.org
Information on this title: www.cambridge.org/9781107605824

© Cambridge University Press 1911

First published 1911
First paperback edition 2011

A catalogue record for this publication is available from the British Library

ISBN 978-1-107-60582-4 Paperback

*With the exception of the coat of arms at
the foot, the design on the title page is a
reproduction of one used by the earliest known
Cambridge printer, John Siberch,* 1521

PREFACE

THE object of this little book is to present a simple account of modern views on the relationships of the chief groups of the animal kingdom, the principal grounds on which the modern classification of these groups may be justified, and an outline of the evidence by which some of the main streams of animal evolution can be traced. The science of comparative anatomy with its theoretical super-structure of phylogeny (the ancestral pedigree of animals) rests so largely upon facts which can only be satisfactorily appreciated by a course of laboratory experience, that it is difficult to present the subject intelligibly to readers who have not already acquired some discipline in zoological science, but, to those who are acquainted with the structure and development of a few types of animals, it may prove instructive to consider some of the wider problems of animal relationship, and to link up their knowledge of special types with a more general appreciation of the probable course organic evolution has taken. In tracing out the course of organic evolution the importance of existing primitive types of animals

has proved very great. Relics of a distant past, the
features of which have been all but effaced by the
passage of time, they preserve for us a record of
bygone phases of existence, and often point us to
some of the sources from which the modern world
of living things has arisen. Although every detail
concerning these primitive animals is of interest to
the naturalist, in a general exposition it is of prime
importance to bring into relief the special features
by which these animals help us to follow the tangled
thread of phylogeny, and in order to do this satisfactorily it is necessary to compare them with their
nearest living and fossil relatives. The chapters in
this book are therefore devoted to the comparative
study of some of the main groups of the animal kingdom; a study in which the consideration of some of
the most striking instances of primitive animals is
involved, though our attention will not be wholly
devoted to them. In this way an idea of the comparative method in morphology (the science of animal
structure) and of some of its principal results may
be gained, though in the short space available it is
impossible to do justice to the great mass of evidence
upon which morphological science is based. One of
the main objects I have borne in mind has been to
distinguish as clearly as possible legitimate from
illegitimate morphological speculations, since the
failure to distinguish between them has been a

fruitful cause of the discredit which in certain minds attaches to speculative morphology. Although the distinction I have attempted to draw may be arbitrary in some cases, my endeavour has been to show that as long as we confine our speculations within the limits of the great animal phyla or groups, comparative morphology has supplied us with a number of securely founded generalizations of real value, but that when we attempt to unravel the inter-relationships of the phyla to one another our speculations are really valueless.

The study of the past is interesting for its own sake, and also for the doubtful rays of prophetic light it may throw into the obscurity of the future. This is my excuse for the last chapter, which is intended to apply some of the principal ideas gathered in the course of our argument to the present condition of mankind, to his future prospects and to the influence which he exerts upon other forms of life.

The sources from which my facts have been gleaned are certainly many, but perhaps the leading ideas have been taken less from books than from the traditional teaching which I have received in the Oxford school of Morphology.

G. S.

NEW COLLEGE, OXFORD.
July 1911.

CONTENTS

CHAPTER I

THE ANIMAL KINGDOM

THE doctrine of descent with modification teaches us that just as the numerous varieties of any of our domestic animals have certainly been derived from one or perhaps a few wild ancestral species, so the various forms of wild animals now existing are the modified descendants of pre-existing forms, which, if we could follow them back into the uttermost recesses of the past, would exhibit less and less diversity of structure. Finally we would arrive at comparatively few and generalized types from which have sprung all the endless forms of animal life in past and present times.

Although in imagination we can retrace this evolution of animal life to its very simplest origins, yet when we come to replace our imaginary scheme by a series of animal types which are known to exist or to have existed in past times, it is found that there are very serious gaps in our knowledge, and that so far from being able to reconstruct the complete ancestral history of the animal kingdom in an objective manner we are forced to accept a great deal of it on trust. The greatest triumphs in this attempt to piece together the actual history of animal change have been won within the last fifty years from the study of vertebrate animals, especially of the most modern and dominant group of them, the Mammalia, and it is no exaggeration to say that we can now trace the history of all the principal kinds of living placental Mammalia through a series of almost continuous gradations back to their common origin in Eocene times. But the Mammalia, geologically speaking, are modern animals, their evolutionary history has been accomplished with comparative rapidity, and their durable fossilized remains have been deposited in strata which in many cases have not been excessively disturbed by subsequent events. We might dwell upon other achievements won by the application of the evolutionary theory to the study of living and extinct animals, but our present object is rather to pass the animal kingdom hurriedly in review[1] and indicate the serious gaps in the evolutionary scheme which there seems little prospect of ever filling in, except by unverifiable conjectures.

The simplest animals known are those whose

[1] In Appendices A and B (pp. 150, 152) a summarized classification of the animal kingdom and a table of the chief geological periods may be found useful for reference.

bodies consist of a single nucleated cell, the *Protozoa*, and these are contrasted to all other animals whose bodies are built up of numerous cells, arranged in definite tissues, the *Metazoa*. Now it is true that certain intermediate forms are known, so-called colonial Protozoa, whose bodies are formed of small colonies of cells, e.g. *Volvox*, but these organisms do not show any affinity to any known Metazoa, their relationship being indeed rather with the more lowly forms of multicellular plants.

The most simply organized of the Metazoa are the *Coelenterata*, including the Jelly-fish, Fresh-water and Marine Polyps, Sea Anemones and Corals, in which the body consists of essentially two cell-layers forming a double cylinder, the one enclosing the other, the outer one, the *ectoderm* being protective in function, the inner, the *endoderm*, enclosing a central cavity and exercising the functions of digestion and absorption (Fig. 1). Between the two layers is a structureless membrane, the *mesoglaea* ; this membrane may be greatly swollen, as in the jelly-fishes, and cellular elements from the ectoderm may invade this jelly and give rise to muscular, nervous, or skeletogenous tissue, but the tissue so formed cannot be regarded as constituting a true cell-layer, as it is a subsequent derivation from the ectoderm and is very frequently absent. The reproductive cells (ova and spermatozoa) may arise in

1—2

these animals either from the ectodermal or the endodermal cell-layers.

In animals of the Coelenterate type the symmetry of the body is radiate; we cannot ordinarily distinguish a dorsal or ventral surface, a right or a left side of the body. The Coelenterates pass their life either fixed to some object from which they stretch out their tentacles in all directions in search of food, or else, as in the case of the jelly-fishes, they float or propel themselves in the water very much at random, being unable to direct their course for any prolonged period in a straight line. The organs of sense and locomotion in these animals are disposed radially, hence the absence of direction in their movements.

When we pass to the next lowest group of animals, the *Platyhelmia* or Flat-worms, we meet with an entirely different type of symmetry, which is functionally correlated with a more advanced and purposive mode of locomotion. The body has ordinarily a flat ventral surface upon which the animal creeps and on which the mouth opens; the opposite surface is the back or dorsal surface, and the animal is bilaterally symmetrical about its long axis. We can distinguish an anterior and a posterior end; at the anterior end the brain is situated and usually, in association with the brain, special sense organs such as eyes are present. The achievement of this bilateral symmetry was evidently a fundamental one for the Metazoa, since it is adhered to with extraordinary uniformity in all the subsequent developments of the animal kingdom, and where it has been lost, as in the Echinoderms or Star-fishes,

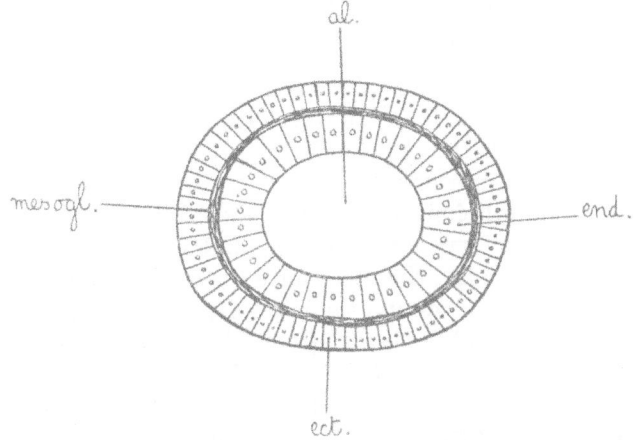

Fig. 1. Transverse section through a Coelenterate. *al.* alimentary canal or gastro-vascular cavity. *end.* endoderm. *ect.* ectoderm. *mesogl.* mesoglaea.

which possess a radial symmetry, the loss is evidently secondary, these animals, as their development shows, being derived from bilaterally symmetrical ancestors. The bilateral symmetry of the Metazoa, from the Platyhelmia upwards, has exercised a profound influence on their evolution. Acquired originally as a necessary condition for movement directed in a straight line it has led to the concentration of sense organs with their nerve-centres in a forwardly directed head, and to the disposition of all the important organs in pairs on either side of the median axis along which the animal moves. In practically all free-living and active animals the bilateral symmetry is retained, and wherever it has been lost, as in many parasitic and fixed forms, we can always perceive that the loss is a result of the lack of necessity of movement.

The Platyhelmia differ from the Coelenterata not only in their type of symmetry, but in possessing a definite cell-layer, the mesoderm, interposed between the outer ectoderm and the inner digestive endoderm. Within this middle layer, from which the muscular and supporting structures are formed, the reproductive cells are always developed as hollow sacs lined with a fairly regular epithelium or investment of cells (Fig. 2).

We do not know by what steps the bilaterally symmetrical Metazoan with three germ-layers and saccular reproductive organs in the mesoderm was derived from the Coelenterate type, though certain Coelenterates have taken to a creeping habit and have developed a bilateral symmetry. These animals are, however, probably highly specialized Coelenterates, and do not represent real intermediate forms, so that we must admit that the transition from the Coelenterate to the Platyhelminth type, if it ever took place, is unrepresented by any living or fossil animal.

We are equally at a loss in trying to trace the origin of the *Nemathelminthes* or Round-worm Phylum[1]. These are three-layered, bilaterally symmetrical, unsegmented round worms, many of them

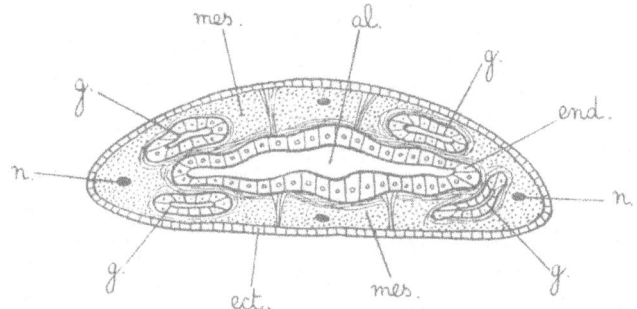

Fig. 2. Transverse section through a Platyhelminth. *al.* alimentary canal. *end.* endoderm. *ect.* ectoderm. *mes.* parenchymatous mesoderm. *g.* reproductive sacs or gonads representing coelom. *n.* nerve-cords.

living in decaying vegetation, others as intestinal parasites in man and Vertebrates (*Ascaris*). They are triploblastic animals like the Platyhelminths, i.e. their tissues are developed from three germ-layers in the embryo, but their internal organs, intestines and

[1] The term Phylum denotes one of the great branches of the animal kingdom (see Appendix A).

reproductive organs, instead of lying in a packing-tissue of mesoderm, are contained in a cavity full of fluid. This cavity represents a blood-system but it is not divided into regular vessels, and no true circulatory system exists as in higher forms.

The Phylum *Nemertea,* containing the Nemertine worms, may perhaps be regarded as affording a link between the Platyhelminth type and the next phylum, the Appendiculata. The Nemertines are round worms which live for the most part a free life in the sea, though a few are found in fresh-waters, a few live on land in damp situations, while a few are parasitic. The Nemertea agree with the Platyhelminths in having a packing-tissue of meso-derm surrounding their internal organs, but in this packing-tissue a definite circulatory system with a heart is differentiated. The body in general is un-segmented as in Platyhelminths, but the reproductive organs consist of laterally paired hollow pouches arranged in linear series along the body. This gives the appearance of metameric segmentation, and it is possible that this represents the beginning of the condition of true metameric segmentation found in the next phylum, the Appendiculata. It must be noted, however, that there is a wide gap between the Nemertea and Appendiculata in the matter of meta-meric segmentation, for whereas in the Nemertea only the reproductive pouches are involved in the

segmentation, in the Appendiculata the muscles and
nervous system have become incorporated in the
segmentally repeated sacs which give rise to the
reproductive organs, and are themselves metamerically segmented. The nervous system of the
Nemertea is also totally different from that of the
Appendiculata, being built on the Platyhelminth
plan, with lateral dorsal and ventral longitudinal
cords, whereas in the Appendiculata there is invariably the double ventral cord with segmental
ganglia upon it.

The institution and vindication of the Phylum
Appendiculata, including a vast quantity of such
various organisms as Annelids or segmented worms,
Centipedes, Scorpions, Spiders, Mites, Crustacea
and Insects, the demonstration that all these widely
different organisms can be proved to exhibit an
essentially similar plan of organization owing to
their having descended from a common ancestor in
the remote past, is one of the most noteworthy
achievements of zoology in recent times, since it
depends entirely on the results of comparative
anatomy and embryology, without the assistance of
any fossilized remains. The key to the situation was
given by the discovery and investigation of a most
interesting animal, *Peripatus*, a true link with the
remote past, which helped to bridge over the gulf
between the segmented Annelid type and the

Arthropod. This question will be considered in more detail in a future chapter.

Now, although the discovery that all these various organisms could be shown to be parts of one intricate evolutionary process from a common type was a satisfactory achievement, yet the origin of the Appendiculate phylum, as indeed of every other phylum, remains shrouded in mystery, save for the possible glimpse of light thrown upon it by the organization of the Nemertea. The great and characteristic acquirements of the Appendiculate phylum, by which it shows an advance of organization over all the lower phyla, are as follows, though we must bear in mind that most of them may be lost in particular cases as the result of degeneration. Firstly, as the name implies, the Appendiculata possess limbs, muscular processes of the body which may become jointed and complicated in structure. Secondly, their bodies are built up of a head and a tail and an intervening region of a varying number of similar rings or segments following one another in linear series. Each one of these segments may carry a complete set of organs, a pair of limbs, a nerve-ganglion, a pair of excretory organs and so forth, so that, discounting the head and the tail, the organism may be said to be built up by the serial repetition of a number of homologous and similar parts. Such an organism, e.g. an Earthworm, is said to be *metamerically segmented.* Thirdly, a characteristic development has occurred in the mesoderm or middle body-layer of these animals. In the phyla so far considered there

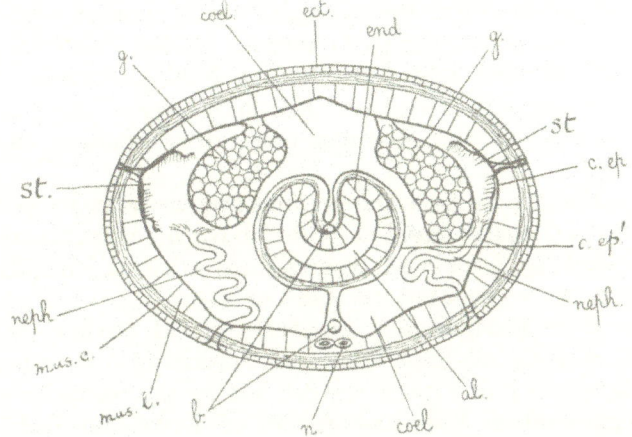

Fig. 3.   Transverse section through an Annelid.   *al.* alimentary canal.   *end.* endoderm.   *ect.* ectoderm.   *coel.* perivisceral coelom.   *c.ep.* coelomic epithelium applied to body wall (somatopleur).   *c.ep'.* coelomic epithelium applied to gut (splanchnopleur).   *b.* blood-vessels.   *n.* ventral nerve-cord.   *neph.* nephridium.   *st.* coelomostome or gonadial funnel.   *g.* gonad.   *mus.c.* circular muscles.   *mus.l.* longitudinal muscles.

has been no formation of definite cavities or spaces in this body-layer, no differentiation of blood-vessels from a body-cavity; but in the Appendiculata a spacious cavity is developed in the mesoblast lined with regular epithelial walls ; this cavity, the *coelom*, forms the general body or perivisceral cavity into which the viscera and other organs are projected (Fig. 3). Besides this cavity, which contains a special fluid of its own, the coelomic fluid, another separate system of canals is formed in the mesoblast, quite distinct and disconnected from the coelom ; these are the blood-vessels, which contain a definite fluid of nutritive and respiratory function, the blood.

Finally, we have the nervous system concentrated into a dorsally situated brain above the mouth, connected by a pair of commissures with a ventrally situated, segmented, double nerve-cord.

Here then, in this immense animal phylum, we meet with at least four characteristics of prime importance, the presence of limbs, the metameric segmentation, the development of a coelomic body-cavity and of a separate vascular system, and the ventral segmented nerve-cord ; but as to the origin of any of these characters we are really in the dark and can only vaguely speculate as to the mode of their acquisition.

The *Mollusca* or Shell-fish constitute another great phylum of the animal kingdom, the component groups of which have been shown by the industry of naturalists to be reducible to a similar plan of organization and therefore to constitute a single evolutionary series, but we have no certain knowledge as to whence the Molluscan type has come or with what other phylum it has any close relationship. At some very remote time we may suppose that the Mollusca had a common origin with the Appendiculata, since a larval form, the Trochosphere, is common to certain members of both phyla. The Mollusca are also coelomate animals like the Appendiculata, but they possess no trace of metameric segmentation, and there is nothing to indicate whether they branched off from the common stock of the Appendiculata before segmentation had been acquired, or whether they have subsequently lost this characteristic.

We may omit certain small and little known phyla, often containing only a few organisms of altogether obscure origin, and pass on to the last remaining important phylum, perhaps the most important of all, as it comprises all the highest forms of animal life, the *Chordata*. The Chordate type, such as is exhibited by any typical Vertebrate animal, for instance a Fish, a Reptile or a Mammal, possesses a number of distinctive features. Like the Appendiculata they are metamerically segmented, but this segmentation, although it is apparent in the embryo, becomes obscured in the adult, which palpably exhibits its segmentation only in the vertebral column or backbone and in the spinal

nerves which issue from the nerve-cord between the
vertebrae. They also resemble the Appendiculata in
the possession of a coelomic body-cavity and a
separate blood vascular system. But the Chordates
differ from the Appendiculata in the fact that the
central nervous system or nerve-cord lies above
instead of below the alimentary canal, and the nerve-
cord itself is hollow, instead of being a double, solid
cord (Fig. 4). Although the Chordates resemble the
Appendiculata, and especially the Annelids, more than
they resemble any other Invertebrate phylum, there
are very grave difficulties in the way of deriving
them from the Appendiculata, a subject which will
be discussed in a future chapter.

In the Chordate phylum, then, despite the fact
that in separate groups of the phylum some of the
most complete and convincing evolutionary series
have been worked out, the origin of the phylum
itself remains one of the most vexed questions of
zoology about which competent zoologists hold the
most diverse and contradictory opinions.

Our short review of the animal kingdom is now
finished, and hurried and utterly inadequate as it
necessarily is, it may have served to indicate that
in the present state of knowledge we are able to
separate the various components of the animal
kingdom into a certain number of divergent groups
or phyla; and that within the limits of these phyla,

although we are far from knowing even approximately in every case the course evolution has

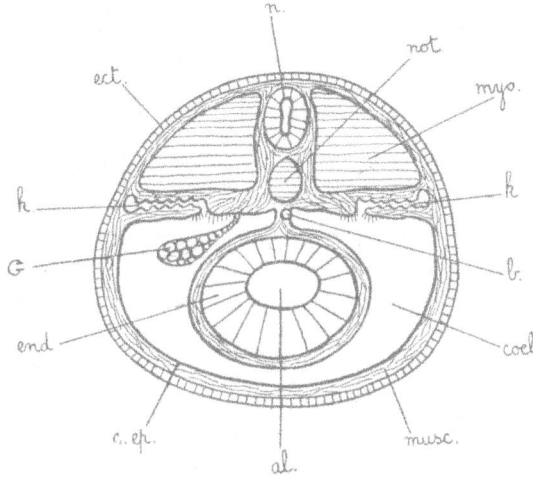

Fig. 4. Transverse section through a young Vertebrate. *al.* alimentary canal. *end.* endoderm. *ect.* ectoderm. *c.ep.* coelomic epithelium (somatopleur), the splanchnopleur is applied to gut and is not lettered. *coel.* coelom. *myo.* dorsal myotom or muscle block. *n.* nerve tube. *not.* notochord. *k.* a kidney tubule or coelomoduct opening into coelom by a ciliated funnel. *G.* gonad, only represented on one side. *b.* dorsal aorta. *musc.* latero-ventral body muscle.

followed, yet we are safe in concluding that all the diverse ramifications have originated in the remote past from some one common source, or type. But

when we attempt to go behind the phyla and discover their origin and inter-relationships, we leave
the firm ground altogether and wander in a slippery
and nebulous region of speculation.

It is true that certain hypotheses of a plausible
character have been suggested which may have
satisfied uncritical minds, and which we often hear
advanced as a part of ascertained science and accepted in an otiose spirit. We are urged to believe
that life "originated" in certain chemical compounds
which on attaining a certain degree of complexity
began to exhibit the fundamental properties of life ;
that from these comparatively structureless masses
the nucleated cell was evolved and the unicellular
animals and plants, the Protozoa and Protophyta,
made their appearance ; that these gave rise to cell
colonies, the Metaphyta on the one hand, and the
Metazoa on the other. The Coelenterate type of
organization is presented to us as the form in which
the early Metazoa had their being, and from this by
the addition of the mesoderm we arrive at the
Platyhelmia, and from them by the addition of a
coelom and of metameric segmentation at the
segmented Annelid. And so by the addition and
subtraction of their characteristic qualities we may
pass in imagination from one fundamental type to
another ; but what is there of reality in these speculations ? They rest not on any objective evidence,

but upon the tendency of the mind to pass from the apparently simple to the manifestly complex, and to regard the former as primitive and ancestral and the latter as secondary and derivative.

The same tendency to regard the simplest organisms within a phylum as the primitive and ancestral forms was exhibited by the earlier morphologists, but a more exacting scrutiny has revealed that almost all these simple animals are degenerate, or simplified, not the retainers of a primaeval simplicity.

We still labour under the old misapprehension as to the age of the earth and of the life upon it. The origins of the animal phyla belong to a past submerged far below even the vast depths of the oldest stratified rocks. It is said with perfect justice that existing animals, including the simplest Protozoa, are only the topmost twigs of the vast genealogical tree of animal life, but it can be said with equal justice that the oldest fossil-bearing rocks do not carry us below the topmost branches. In the oldest fossiliferous rocks we meet not only with typical representatives of the Crustacea, a specialized branch of the Appendiculate phylum, but with actual families of Crustacea (Cypridae, Nebaliidae), which exist at the present day, while in the next oldest strata are fishes, representatives of the Vertebrata, the highest type of organization in the animal kingdom. Much as the geological record teaches us and may yet teach us, it does not seem to touch the origins of the animal phyla. The strata in which those archives were deposited, if they ever were deposited, have been crushed and burnt and re-crystallized out of all recognition and their secret is lost perhaps for ever, and we cannot even guess at the nature of it.

In what sense, then, can we speak of existing animals as primitive or as affording links with the past? In the following chapters we propose to deal with three classes or types of facts which throw a light upon the past history of life. In the first place there are animals still in existence which, although clearly belonging to some recognized animal group, have not advanced to the same stage of specialization in all directions as their related forms, but have retained in many respects a more generalized plan of structure; they have, so to speak, stood still on the evolutionary road while the main troop of their kind has travelled on and ultimately diverged into many and various paths. In many cases we can trace the geological history of such animals and we find them existing at some very remote epoch practically in the identical form which they exhibit to-day. Why it is that certain animal types should remain so constant for such long periods, while others related to them should undergo the most fundamental changes, we can only guess at. To find the reason is rather the province of the student of variation and heredity; it is our task to show that it is so. The special importance which attaches to such animals is that they enable us to see relationships between modern groups which have undergone rapid and far-reaching changes, and whose connection with one another and with other forms would be obscured were it not for the existence of these ancient, connectant or primitive types.

We shall also deal shortly with the young developmental stages of certain animals, which as adults do not particularly betray their affinities, but which during development pass through stages of the greatest significance in revealing their nature and relationship.

Finally our attention may be attracted to animals that are on the threshold of, or have recently suffered, extinction, and here the link is not so much with the ancient past of animal life, as with that more tangible past when man had already begun to exercise dominion over the earth and to exert that immense influence over living creation the final effects of which are still to be witnessed.

CHAPTER II

THE SIMPLEST ANIMALS AND PLANTS AND THE ORIGIN OF LIFE

LIVING substance, or the matter in which the phenomena of life are manifested, possesses a very highly complex physical and chemical structure, but this complexity is due not to the number and variety of chemical elements which enter into its composition, but to the almost infinite number of combinations which these elements are capable of forming among themselves. The largest and most complex molecules which enter into the composition of living matter, and upon which many of the most essential vital properties are supposed to depend, are the proteids, and these bodies are made up of the elements carbon, hydrogen, oxygen, nitrogen, sulphur and phosphorus. The continuity of life depends upon the possibility of bringing a continual supply of these elements to living matter, and of recombining them into the living substance after they have passed out into the dead world. Living matter is continually wasting away, and its elements are dispersed both during life and at death as inorganic material; but the waste is repaired by the capacity of living matter to assimilate to itself this dead matter and to rebuild its body from it. In this process of reclaiming the elements necessary for life after they have passed out of the living stream, plants play by far the most important part, since they alone are capable of building up their bodies from simple inorganic substances, whereas animals can only use as food substances which have been already elaborated into complex organic bodies by the action of life. The food-stuffs of an animal from which it obtains its carbon, nitrogen, oxygen, hydrogen, sulphur and phosphorus, are fats and carbohydrates (which contain the elements C, H and O) and proteids (which contain besides C, H and O, the elements N, S and P).

Besides these substances an animal requires free oxygen, since its energy is derived by the combustion of the carbon in its food-stuffs; this oxygen is obtained from the atmosphere or from solution in water; and in addition water and certain mineral salts are necessary. Now, the food-stuffs of a green plant are entirely different; it does not make use of any organic material, but obtains its carbon from the carbonic acid gas (CO_2) present in the atmosphere, from which it abstracts the carbon through the aid of its chlorophyll acting in the presence of sunlight;

it obtains its nitrogen from the mineral nitrates in the soil, its hydrogen from water, its oxygen from the atmosphere and from water, and its phosphorus and sulphur from salts in the soil. The green plant, therefore, is the means by which the elements necessary for life, after they have broken away from the living bodies of animals and plants, are brought back again and put into circulation in the stream of life ; and if it were not for the existence of green plants all life on the earth would very soon come to an end.

Animals are therefore absolutely dependent in the end on the existence of green plants, but animals in their turn, although not necessary for the life of plants, form a very large part of the raw material from which plants draw their sustenance. Thus animals as the result of their activity breathe out CO_2 into the atmosphere, which supplies the plants with their carbon. The excreta and dead bodies of animals manure the ground for plants, i.e. decompose into substances which ultimately yield the nitrates, phosphates and sulphates from which the plant builds up its body.

In this manner there is a continual circulation of the chemical elements necessary for life from plant to animal and animal to plant ; but although the animal ultimately yields up its substance to the plant, yet the existence of animals is in no way so necessary for the plant as the existence of the latter is for the animal. Plants could exist perfectly well without animals, since their own respiration and the decay of their own bodies would furnish them with the necessary materials of life ; but if all the plants were destroyed and only animals left existing, they might subsist on one another for a time, but their substance would rapidly leak away as inorganic matter, which they would have no power to reclaim and assimilate.

In the process of the circulation of the chemical elements necessary for life, an essential *rôle* is played by the organisms known as Bacteria, without whose activity life as at present constituted would very rapidly come to a standstill and soon perish altogether. We have seen that the carbon necessary for life is reclaimed from the dead world by the green plant, which through the aid of its chlorophyll in the presence of sunlight is able to abstract the carbon from the carbon dioxide gas of the atmosphere and incorporate it with its own substance. In the case of nitrogen, the green plant is able to abstract this element from one source and from one source only, namely, the inorganic mineral nitrates in the soil.

Now, the nitrates in the soil are almost entirely derived from the decay of organic matter—the dead bodies and excreta of animals and plants—but the production of nitrates from this organic material is a complicated process brought about by the concerted action of several kinds of bacteria. When an organism dies, be it animal or plant, its body cumbers the ground, and is at once the seat of fermentative processes of decay, brought about by the putrefactive bacteria, which break up the complicated proteid materials of the body into simpler nitrogenous substances and finally into ammonia compounds. When the decay is complete other bacteria attack the ammonia compounds and form mineral nitrites from them, and by a further process of oxidation a new set of bacteria convert the nitrites into nitrates, and thus at last the nitrogen has reached the form in which it can again be taken up by the green plant and reassimilated into a living body. Each step in this process is the work of a special kind of bacteria, and were it not for the activity of these organisms, the death of every living body would mean a total loss of so much nitrogen to life, so that in the end life would cease from nitrogen-bankruptcy.

Besides the power of getting at the nitrogen in dead bodies and converting it finally into nitrates, certain bacteria have the unique power of fixing atmospheric nitrogen, thus actually increasing the total available nitrogen for the use of life. These bacteria play an important part in agriculture, as they enrich the soil in which they grow without themselves being dependent on the presence of any nitrogenous bodies in solution.

Bacteria are therefore to the green plant in respect to nitrogen what the green plant is to all other organisms in respect to carbon. Without the assimilatory power of chlorophyll life would cease from carbon-bankruptcy; without the putrefactive and nitrifying power of bacteria life would soon cease from nitrogen-starvation.

These elementary but fundamental facts of plant and animal assimilation have been shortly outlined here, because the question of the origin of life, obscure as it is, would seem to centre principally round the problem of assimilation, that power by which life selects the elements necessary for its own substance from the dead compounds around it, and builds them up into its own integral body.

It cannot be said that the progress of biology has thrown any light on the problem of the origin of life. Indeed, in former times, before the researches of Redi, Needham and others had culminated in the classical work of Pasteur, the spontaneous generation of life from dead matter was generally held to be a process of constant and everyday occurrence, so that the origin of life might seem at any rate a soluble problem. But now, when the doctrine of spontaneous generation has been finally demolished and the contrary doctrine that all life proceeds from pre-existing life is universally accepted, the origin of life is necessarily shifted back to a remote and intangible past when the conditions of existence were unimaginably different from those at the present time.

It is very difficult to believe that life originated in anything like the form presented to us to-day by the simplest living organisms known, namely the Protozoa, Protophyta and Bacteria. It may be true that these simple unicellular organisms are primitive, in the sense that they represent a very early stage in animal evolution when the animal or plant body was composed of a single cell, but this early stage is very far removed from the origin of life itself. The Protozoa comprise a great variety of animal forms, characterized by comparative simplicity of structure and by the fact of their body being composed of a single cell. Such a Protozoon as *Amoeba* (Fig. 5 A), which consists of a small mass of naked foam-like protoplasm in which a nucleus of more solid consistence is differentiated, represents the lowest grade of animal organization known to us. The whole surface of the body is capable of protrusion for the purposes of locomotion or for engulfing food; an excretory vacuole can be formed at any part of the body, and reproduction is effected simply by the division of the body into two, the division of the

nucleus preceding that of the cell-body or cytoplasm.
Such a cell, consisting of nucleus and cytoplasm, is
the lowest unit of living matter, and all the supposed

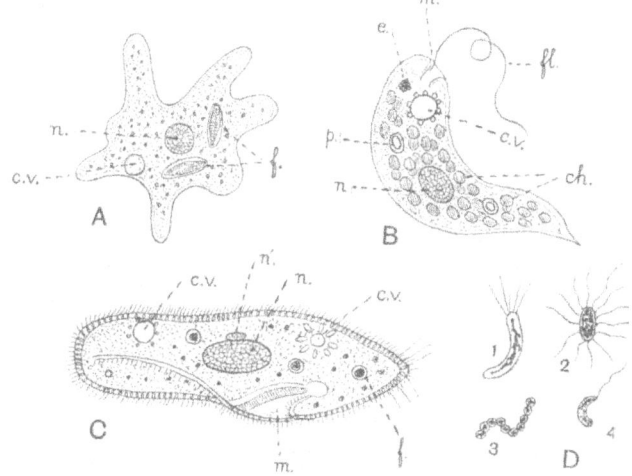

Fig. 5. A. *Amoeba proteus.* B. *Euglena viridis.* C. *Paramecium
caudatum.* D. A group of Bacteria. 1, Spirillum. 2, Bacillus.
3, Streptococcus. 4, Vibrio. *n.* nucleus. *n'.* micro-nucleus.
m. mouth. *c.v.* contractile vacuole. *f.* food vacuole. *e.* eye
spot. *ch.* chloroplastid. *p.* pyrenoid with starch concretion.
fl. flagellum. In C the small bodies in the cuticle represent
trichocysts, some of which are discharged in the anterior region.

cases of living organisms in which a nucleus was
thought to be absent have been proved to be in-
correct.   In the Bacteria (Fig. 5 D), it is true, the

nuclear material is not sharply divided off from the
cytoplasm by a definite membrane, but granules of
nuclear material are nevertheless always present and
can be readily recognized by their characteristic
staining properties.

The Protozoa show a considerable range of organization, from an almost structureless form like that
of *Amoeba* to the comparative complication of some
of the Ciliate Infusoria (Fig. 5 c), which may possess a
definite mouth, cuticle, excretory vacuole and stinging
threads all within the limits of a single cell.   Other
Protozoa may attain a very large size and in these
cases a great multiplication of nuclei may take place,
so that the organism may be multinuclear though
not truly multicellular.   Again, in many Protozoa,
in addition to the simple mode of reproduction by
division, a complicated sexual phase may occur in
which the individuals are differentiated into male
and female gametes in every way comparable to the
spermatozoa and ova of the higher Metazoa.   Therefore although it may be true that the Protozoa are
the simplest known organisms, yet they exhibit all
the characteristic phenomena of the higher animals
and do absolutely nothing to bridge the gulf between
the living and the non-living.   Nor in the all-im-
portant question of their mode of nutrition do they
differ essentially from the higher animals.   The Protozoa, including the simplest of them such as *Amoeba*,

are absolutely dependent on being supplied with complex food materials already elaborated by the activity of plant or animal life. The *Amoeba* feeds on minute plants, such as Diatoms, on Bacteria, and to some extent on other animals, and it could not possibly have existed before a plentiful supply of plant life had covered the earth. The same is the case with all other Protozoa. It is therefore futile to look for the beginnings of life among the carnivorous and herbivorous Protozoa, or anything nearly related to them, as their whole structure and way of life is arranged on the assumption that living material already exists to furnish them with food.

It seems equally futile to regard the Bacteria as representing the primal forms in which life appeared. It is true that these organisms stand in a much closer relation to inorganic nature than do the Protozoa, since they are able to obtain their nitrogen not only from inorganic compounds but also in some cases from the pure gas in a free condition, and we may remember the remarkable powers possessed by the sulphur and iron bacteria of obtaining energy by the oxidation of simple inorganic iron and sulphur compounds. But the Bacteria are helpless in another direction; since they are without chlorophyll they cannot obtain their necessary carbon in an inorganic form[1], and hence they are obliged to seek this element

[1] This has been disputed in a few doubtful cases.

among organic carbon compounds which have been elaborated by the action of other animals or plants. This is the reason why bacteria always congregate in decaying animal and vegetable matter, and it would seem that their existence presupposes the existence of other forms of life to supply them with the requisite combined carbon.

We are left therefore with the chlorophyll-possessing green plant as the conceivable starting point for living matter. From the point of view of the problem of assimilation, the green plant is in much the most independent position, as it can form its own substance from inorganic matter without the aid of other living beings. It is indeed difficult to escape from the conclusion that the earliest living organisms possessed chlorophyll, as, without the power of combining inorganic carbon, the continued growth of living substance seems, as far as our experience goes, to be an impossibility.

There is a large group of unicellular organisms now in existence, the Flagellata, which propel themselves through the water by means of one or more whip-like cilia, some of which are without chlorophyll and must be considered as animals, since they nourish themselves on already elaborated organic matter while others possess chlorophyll and have the power like any green plant, of nourishing themselves on carbon dioxide and purely inorganic compounds.

A good example of these Flagellata, which can with difficulty be classified either as an animal or a plant, is *Euglena viridis* (Fig. 5 B), a very common organism, which is often responsible for the bright green scum that collects on the surface of puddles and ponds. Such a green scum is often found to consist of a mass of small green organisms which wriggle and swim in the water by means of a flagellum and by the contractions of the elastic body-wall. *Euglena viridis* contains chlorophyll by means of which it can assimilate carbon dioxide (CO_2), and accumulations of starch-like substances are found in the body. It can flourish on purely inorganic substances like a green plant, but its animal properties are shown in its possession of a mouth, its capacity for assimilating organic food, and in the fact that it possesses a pigment spot sensitive to light. Though we may admit that the plant nature of *Euglena* preponderates over the animal nature, yet there are close relations of *Euglena* among the Flagellata which are complete animals in their manner of nutrition, and which no one would hesitate to classify as animals. The Flagellata as a group can, therefore, be confidently ascribed neither to the animal nor to the plant kingdom. In their mobility, the absence of a cellulose cell-wall which is present in all undoubted plants, and in the peculiar elasticity with which they assimilate now in the

manner of animals and now in the manner of
plants, they seem to constitute a connecting link
between animals and plants and to stand at any
rate near to the branching of the road which led to
the great dichotomy of all living things. That the
Flagellata are truly in this position is supported
by the widespread occurrence of a flagellated phase in
the life-history of so many unicellular animals and
plants, as well as in the spermatozoa of practically
all the Metazoa and of the more primitive Metaphyta.

We set out from the standpoint that any theory
of the origin of life should take account of the fundamental facts of assimilation, as lying at the very
basis of all vital processes. If we assume that life
at its origin had any of the properties of the simplest
forms of life as they now exist, we can hardly escape
from the conclusion that the presence of chlorophyll
was the necessary precursor of life. We must of
course admit that chlorophyll in the present state of
nature is only produced by life, so that the origin of
the chlorophyll itself is a problem as cogent and as
obscure as that of life. If however we prefer to do
without chlorophyll, and conceive of the first living
things as existing without this substance, then we are
supposing the existence of a state of life about which
we can predicate nothing, the continuance of which
must have depended on entirely unknown properties,
for the ascertainment of which we have at present no

data, or methods of enquiry. This being the state of this obscure problem, we can hardly blame the genial physicist who put forward the hypothesis that life never originated on our planet but arrived fully formed in a meteorite, nor need we be surprised that this hypothesis has been found capable of extension into the doctrine that life is coeval with matter and that living germs are constantly being propelled through space by radiation to find a resting-place on whatever planet has reached a condition suitable to their existence and propagation. We know of bacterial spores that could withstand the intense heat and cold and lack of oxygen of interstellar space, but a serious objection is the sensitiveness of such spores to ultraviolet light, to the intense action of which they would be exposed in their passage. Enough, however, has been said to show that the problem of the origin of life, whilst raising many subsidiary questions of scientific interest, is hardly itself within the range of serious scientific treatment.

CHAPTER III

THE APPENDICULATE PHYLUM

I℉ we examine an ordinary Earthworm we may at first be at a loss to understand how it is rightly included in the Appendiculate phylum, since we can at first detect no evidence of the presence of limbs. But if the lower surface of the body is gently rubbed with a finger, a certain roughness will be felt due to the presence of groups of bristles partially embedded in the body-wall of each segment. It is by means of these bristles or setae that the worm obtains a purchase on the earth for its movements, and they indicate the degenerate remains of limbs. The marine relations of the Earthworm, such as may sometimes be seen swimming with graceful, sinuous movements in rock-pools at low tide, possess much better developed limbs which project from the sides of the body on each segment and are armed with longer and more conspicuous bristles than those of the Earthworm. All Annelids, then, possess limbs or the rudiments of limbs, and in this respect they resemble such animals as Centipedes, Insects, Spiders,

and Crustacea[1]. They further resemble these animals
in having the body composed of a number of segments, marked externally by rings; they are all
metamerically segmented animals.   But if we open
any segmented worm by an incision along the middle
dorsal line, and compare what we see with an Insect
or Crustacean treated in a similar way, very great
and fundamental differences will be apparent.   In an
Annelid such as an Earthworm we find that we have
cut into a body-cavity which is divided into as many
partitions as there are external rings or segments,
and in this segmented body-cavity the various organs,
such as the alimentary canal, the nerve-cord, the
excretory and reproductive organs lie.   The body-
cavity or coelom contains a colourless fluid, the
coelomic fluid, in which amoeboid corpuscles float,
and it has no connection at all with the blood vascular
system which will be at once distinguished as a
number of vessels of various sizes containing a red
fluid, the blood, and distributed all over the body.
The blood-vessels form a completely closed circuit
in which the blood circulates, and nowhere does this
system communicate with the coelom or body-cavity.

It has been stated that the various organs lie in
the body-cavity.   This statement is not strictly true.
If reference be made to Fig. 3, illustrating a transverse
section through the body of a worm, it will be seen that

[1] For classification of Appendiculata see Appendix A, pp. 150–1.

the alimentary canal, blood-vessels and nerve-cord really lie outside the coelomic cavity, being covered over with a layer of coelomic epithelium. They project into the coelom, but they do not lie in its cavity.

There are however two sets of organs which actually lie in the coelomic cavity, and always bear a very intimate relation to it. These are the excretory and reproductive organs respectively.

The reproductive cells, ova and spermatozoa, are derived actually from the coelomic wall, and the reproductive organs, ovaries and testes, are simply specialized parts of the coelomic cavity engaged in the elaboration of the sexual products. Intimately related with these organs, funnels may be developed connected with a coiled duct leading to the exterior. These are the genital funnels for conveying the sexual products to the outside of the body, and since they too are developed from the coelomic wall they are called coelomoducts (Fig. 3 *st.*). In some worms a pair of these coelomoducts and reproductive organs may be developed on a very large number of segments, but in other cases, as in the Earthworm, they may be restricted to a few segments.

The excretory organs in the Earthworm consist of a pair of coiled tubes in each segment, each with a ciliated funnel opening into the coelomic cavity, and a duct opening to the exterior (Fig. 3 *neph.*). These are the nephridia which collect the waste material of the body from the coelom and pass it to the exterior. Although they are now in very intimate connection with the coelom they are not developed from its walls: they arise in the skin of the embryo and grow inwards into the coelomic cavity, being thus developed in exactly the opposite way to the coelomoducts which are developed from the coelomic walls and grow outwards to the skin.

Now it happens that although the nephridia are the true excretory organs of the Annelid, yet in these and in other animals the coelomoducts, which are originally genital ducts, may lose their primitive function and take on the function of excretory organs. In animals, therefore, we have to distinguish between these two types of excretory organs which may completely resemble one another in structure, the *nephridia* or primitive excretory organs which are developed from the exterior and push into the coelom, and the *coelomoducts*, which are primitively genital ducts and secondarily excretory organs, which develop from the coelomic wall and push to the exterior.

Let us now turn to our comparison with the Arthropod, Insect, Arachnid or Crustacean, which we have dissected. On opening such an animal we shall at once perceive that there is no spacious segmented body-cavity, but the various organs lie in an irregular system of spaces filled with a colourless or faintly blue-coloured blood, the blue colour being due to the presence of a respiratory substance, haemocyanin, resembling haemoglobin but having the iron replaced by copper. There is no distinction in these animals between the body-cavity and the blood vascular system, the body-cavity being, in fact, not a coelom, but formed from greatly swollen veins which surround all the organs of the body. What then has happened to the true coelom? We have seen that in the coelomate Annelid two systems of organs are always intimately related, both in function and derivation, with the coelom, viz. the reproductive and excretory organs : it is therefore natural to look for the remains of the coelom of the Arthropods in connection with these organs. The reproductive organs of the Arthropods are hollow sacs derived from the mesoderm with ducts leading to the exterior, and on theoretical grounds we may suppose that these sacs represent a part of the original coelom. Excretory organs of various kinds are found in the Arthropods, but they are not segmentally arranged ; we may mention the shell and green glands found in Crustacea, coiled tubes which end internally in hollow closed sacs of mesodermal origin. It has been suggested that these closed sacs represent a portion of the coelom, and that the coiled tubes leading away from them to the exterior are coelomoducts. Somewhat similar organs are found in Arachnids. In the

Centipedes and Insects a new type of excretory organ is met with, the Malpighian tubules ; these are numerous tubes which open into the hinder part of the alimentary canal, and though it has been suggested that they represent nephridia, it seems more probable that they are entirely new and independent excretory organs, and that the true nephridia and coelomoducts have been lost together with the portion of coelom into which they originally opened.

The fundamental difference in the relations of the body-cavity and blood vascular system between the Annelids on the one hand and the Arthropoda on the other, is clearly a difference of great physiological and morphological importance. The physiological reason for the disappearance of the coelomic body-cavity in the Arthropoda and its retention in the Annelids and Vertebrata has never been properly worked out. The only explanation we are prepared to offer is that in soft-bodied animals like worms and vertebrates, where the skeleton, if present, is situated internally, the coelom with its fibrous and muscular walls exercises an important mechanical function in protecting the internal organs from rupture by external pressure or by the strain of the muscular system upon the skeleton. The internal organs and especially the intestines are thus protected by being enclosed in a double fibrous bag, the coelomic fluid between the two layers of the bag acting as a lubricating oil.   In the Arthropoda the body is protected from external agencies by the hard external skeleton to which the muscles are attached ; the internal organs are therefore protected from external agencies and from the strain of muscular contraction, and do not require the protective body-cavity.

The presence of a coelom system distinct from the vascular system may also be of importance in resisting the infection of micro-organisms from the alimentary canal, since the coelomic fluid is full of amoeboid corpuscles whose function is to attack and devour any bacteria or poisonous organisms which may find their way into the coelom.

If it is true that the Arthropods are derived from Annelid-like ancestors, we are in need of some definite evidence to support the theory that the coelom in the Arthropods has been displaced by the swelling of the veins, and that the remains of it are to be sought in the reproductive and certain excretory organs.

It is here that a particular living animal gives us the clue we want, a living animal that has survived from a distant age and has retained certain primitive features of organization which help us to piece together the course of events leading to the disappearance of the coelom in the higher Arthropods. The PERIPATUS (Fig. 6), for so the animal is called, is represented by a small number of closely related species which live in various parts of the tropics and of the southern hemisphere, but do not occur in the northern hemisphere. They are found in S. America, in S. Africa on Table-Mountain, in Tasmania and Victoria, in New Guinea and the West Indian Islands. The animal most nearly resembles a Centipede with short and few legs, and its habit is to frequent forests and undergrowth, living under decaying tree stumps and fallen timber. The skin has a velvety appearance;

Fig. 6.   *Peripatus capensis*, drawn from life.   Life size.
(After Sedgwick.)

the movements are slow but not ungraceful and the creature feels its way about with the help of a pair of short antennae, the eyes being ill-formed and built on a simple plan. The eyes in fact resemble more closely the eyes of a worm than the compound faceted eyes found in Arthropods.

In its internal anatomy *Peripatus* at first sight would appear to be a typical Arthropod; there is no internal segmentation apparent, and the body-cavity consists of a number of vascular spaces containing blood. The heart is perforated by ostia as in an ordinary Arthropod, and the reproductive organs are built on the complicated Arthropodan plan. *But in every segment of the body, except the few extreme anterior and posterior segments, there is present a pair of excretory organs, just as in a typical Annelid.* These excretory organs are small coiled tubes which open by ciliated funnels internally into small dilated end-sacs, and they open to the exterior at the bases of the legs.

We must, however, trace the development of these organs in order to appreciate their full significance for the solution of our problem (see Fig. 7).

Peripatus develops directly from the egg without passing through any larval stage, and in a very young embryo soon after the three primary layers of cells have been formed, viz. the ectoderm, endoderm and mesoderm, we find that the mesoderm acquires a cavity, a true coelom, one wall of which is applied to the ectoderm and the other to the endoderm (Fig. 7A). The mesoderm, further, becomes segmented, so that its cavity, the coelom, is cut up into a number of partitions separated from one another by transverse septa, and in this stage it exactly resembles a young worm. The subsequent fate of these coelomic partitions is of fundamental importance. Each coelomic partition divides into

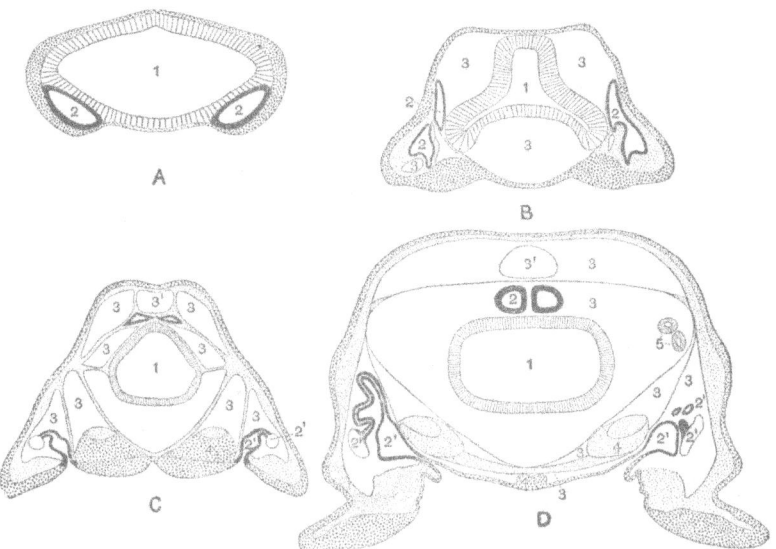

Fig. 7. A series of diagrams of transverse sections through *Peripatus*
embryos to show the relations of the coelom at successive stages.
(After Sedgwick.) A. Early stage. 1, gut; 2, mesoblastic
segment; no trace of the vascular space; endoderm and ectoderm
in contact. B. Endoderm has separated from the dorsal and
ventral ectoderm. The segment is represented as having divided
on the left side into a dorsal and ventral portion. 1, gut;
2, somite; 3, haemocoele. C. The haemocoele (3) has become
divided up into a number of spaces, the arrangement of which is
unimportant. The dorsal part of each somite has travelled
dorsalwards, and now constitutes a small space (triangular in
section) just dorsal to the gut. The ventral portion (2′) has
assumed a tubular character, and has acquired an external opening. The internal vesicle is already indicated, and is shown in
the diagram by the thinner black line. 1, gut; 2′, nephridial
part of coelom; 3, haemocoele; 3′, part of haemocoele, which
will form the heart—the part of the haemocoele on each side of
this will form the pericardium; 4, nerve-cord. D represents the
conditions at the time of birth; numbers as in C, except 5, slime
glands. The coelom is represented as surrounded by a thick
black line, except in the part which forms the internal vesicle of
the nephridium.

two, and of the two cavities so formed on each side in every segment, the one travels up on to the back above the intestine, the other remains below (Fig. 7 B, C). As this division and migration occur, the ectoderm and endoderm draw apart, leaving wide spaces between them, and these are gradually converted into the heart and the blood vascular system of the adult which usurps the function of a body-cavity. The true coelomic partitions on the other hand, the divisions and behaviour of which we have mentioned, do not enlarge to form the body-cavity as in a worm. They remain comparatively small, and the pieces that have travelled dorsally form the reproductive organs, while those that have remained in a ventral position acquire openings to the exterior and are converted into the excretory organs, which are, therefore, coelomoducts (Fig. 7 D). The development of *Peripatus* proves that the body-cavity of this animal is not a true coelom, but is part of the vascular system, and it also proves that a true segmented coelom is formed in the embryo just as in a worm, the remains of which in the adult are to be found in the reproductive organs and the segmentally arranged excretory organs.

Peripatus, therefore, is a real link between the Annelid and Arthropodan type as regards this most important feature, the relation of the body-cavity and vascular system. In all its other characters, perhaps with the exception of the eyes, it is an Arthropod, and approaches most nearly the Centipede or Myriapod type of organization.

Although we are thus able in some measure to link up the Annelids and Arthropods together, the various classes of Arthropods stand in a rather isolated position with regard to one another. We may derive the Myriapods from a Peripatus-like ancestor, and the immense class of Insecta probably comes from a similar source. Like *Peripatus* these animals are all typically air-breathing forms, living on dry land and taking in the air by means of openings on the surface of the body (stigmata) which lead into tubes or *tracheae* distributed to all parts of the body. The other two great classes of Arthropods, the Arachnida and Crustacea, stand in a doubtful relation to the typical Tracheates mentioned above and to one another. The modern and most highly developed Arachnida, viz. the Spiders, are true land-animals and they resemble Insects in possessing tracheae, but it seems certain that the Spiders' tracheae are an independent acquisition and not a legacy from a common ancestor of the Insects. For we know of certain very remarkable Arachnida, of a much more primitive general structure than Spiders, which inhabited the ancient seas and breathed by gills very much in the manner of Crustacea, and these animals have left a descendant, a link with the past, which still exists on the Pacific Coasts of America and Japan. This animal, the KING-CRAB or LIMULUS (Fig. 8), owing to its outward appearance and to its being an inhabitant of the sea, was at first considered to be a Crustacean, but an examination of its anatomy has shown it to be undoubtedly an Arachnid. It grows to a large size, nearly two feet in length, but Palaeozoic fossils (Eurypterida), related to the King-Crab, attained the immense length of six feet. These creatures, looking like gigantic scorpions, apparently lived at slight depths near the sea-shores like *Limulus*, but what their prey was or for what purpose they attained their immense size is not known. That the King-Crab is an Arachnid is conclusively shown by the structure of its eyes and by that of the anterior pair of limbs which are not antennae but pincers as in the Spiders, but the gills, which are carried externally upon the posterior limbs, are folded vascular plates which recall to some extent the gills of Crustacea. The terrestrial air-breathing Arachnid most nearly related to the King-Crab is the Scorpion, and this animal has peculiar respiratory organs, known as lung-books, which are situated on the same segments as the gills of *Limulus*, but they are sunk below the skin and open to the exterior by small orifices, the stigmata. It is practically certain that the Spiders have acquired their chief breathing organs from such lung-books, in

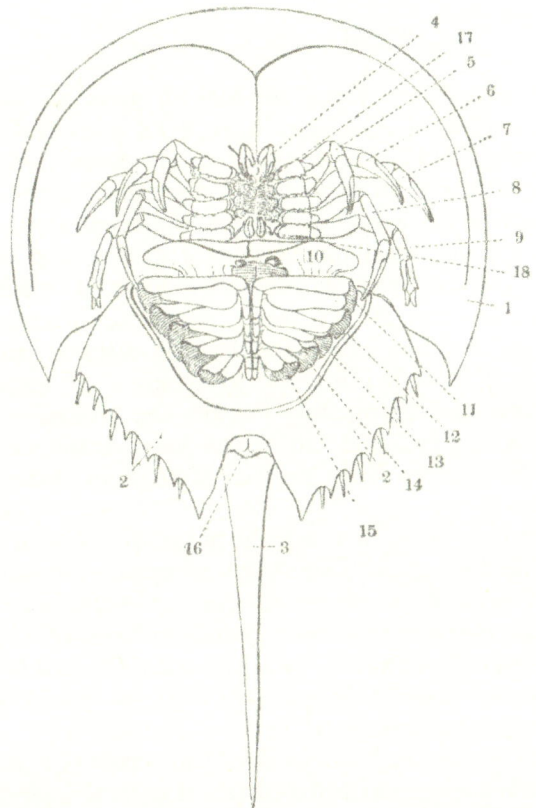

Fig. 8.   Ventral view of the King-Crab, *Limulus polyphemus*, × ½.
(From Shipley and MacBride.)   1, Carapace covering prosoma;
2, meso- and meta-soma; 3, telson; 4, chelicera; 5, pedipalp;
6, 7, 8, 9, 3rd to 6th appendages, ambulatory limbs; 10, genital
operculum turned forward to show the genital apertures; 11, 12,
13, 14, 15, appendages bearing gill-books; 16, anus; 17, mouth;
18, chilaria.

which case they must be looked upon as highly
metamorphosed gills that have been retracted into
the body, and have nothing to do with the true
Insect tracheae in derivation. Certain Spiders (the
Dipneumonous Spiders), besides the respiratory sacs
which are probably homologous with the lung-books
of Scorpions, possess medially situated tracheae
which have had an entirely separate origin, namely
from the sinking below the skin of certain tendons
originally intended for the attachment of muscles.
The Arachnid respiratory organs are therefore not
only entirely independent of the Insect tracheae, but
have themselves been derived from more than one
source.

We have seen, then, within the Arthropod phylum
that the two great terrestrial groups of the Insects
and Arachnids stand apart from one another, and
that the more ancient and primitive forms of Arachnids, so far from affording a link with *Peripatus* and
Insects, are marine animals showing no resemblance
to these latter groups.

The same may be said of the class Crustacea.
They too occupy an isolated position in the phylum,
and we know of no transitional forms between the
most ancient Crustacea and *Peripatus*, Insects or
Arachnids. Here again, as in the case of the
Arachnids, we have an abundantly represented, well-
preserved but very ancient fossil group, the Trilobites

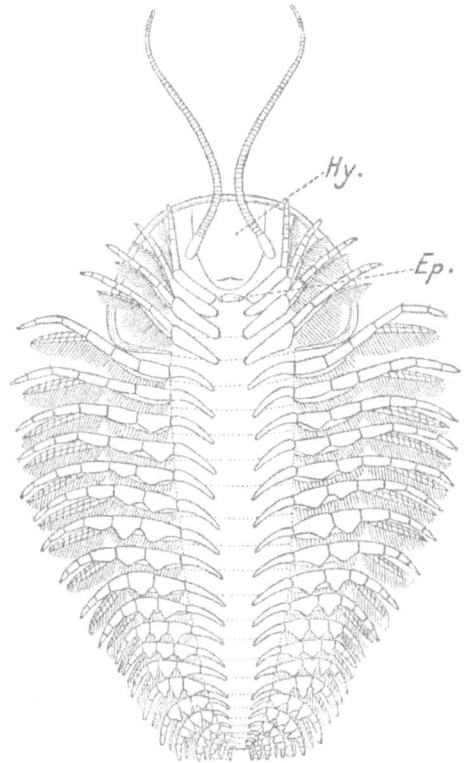

Fig. 9.   *Triarthrus becki*, Green,   × 2½.   Utica Slate (Ordovician), near Rome, New York.   Ventral surface with appendages ; *Ep.* metastome ; *Hy.* hypostome.

(Fig. 9), found in the oldest stratified rocks, which undoubtedly represent very generalized and primitive Crustacea, but they do not form a transition into any of the other Arthropodan classes. It is only of recent years that the relationship of the Trilobites has been definitely established, but the discovery that the anterior head appendages were antenniform, and the succeeding limbs biramous, i.e. consisting of a basal portion carrying two terminal branches, placed their Crustacean affinity beyond doubt. The possession of the biramous limb is the most characteristic feature of the Crustacean class; despite the wonderful variety exhibited by Crustacean limbs, from the pincers of the crab or lobster to the complicated leaf-like paddles of the water-flea or brine-shrimp, from the tooth-like mandible to the tactile second antennae, all the appendages of the Crustacea can be reduced to the biramous plan, with the exception of the first pair of antennae, and it is interesting to know that even in the Trilobites, where all the limbs are in their most primitive and least specialized condition, the first antennae are *sui generis* and of a simple uniramous structure. It may be truly said that the recent discovery of the nature of the appendages in the Trilobites is a powerful vindication of the comparative methods of morphology, so far as the Crustacean limb is concerned.

Within the Crustacean class we meet with two

living animals, *Nebalia* and *Anaspides*, of especial
interest for us, as affording links with the ancient
past, and throwing some light on the course of
Crustacean evolution. The Crustacea fall into two
sub-classes, the Entomostraca, containing small animals
with an indefinite number of segments (from eight to
over a hundred), and the Malacostraca, including the
larger and more highly organized forms, in which the
number of segments, not counting the tail-piece, is
always nineteen. *Nebalia*, a small marine animal of
world-wide distribution, usually living on the sea-
shore at low-tide marks under stones or in the sand,
exhibits a generalized organization which might allow
of its inclusion in either of the two Crustacean
groups; it is neither decidedly Entomostracan nor
Malacostracan, but it exhibits the typical number of
Malacostracan segments, except that an additional
segment is present in the tail region. The limbs,
which are all clearly of the biramous plan, and
the general anatomy, indicate a primitive and un-
specialized condition; the possession of two pairs of
excretory organs in the adult opening on the second
antennae and the second maxillae respectively is of
especial interest, since both these organs do not
persist in the adult of any other Crustacean, though
every Crustacean possesses either one or the other at
some period of its life. In fact *Nebalia* must be
looked upon as a Malacostracan which has retained

4—2

numerous primitive features, some of which it possesses in common with the Entomostraca. It is interesting to observe that *Nebalia*, or very closely allied genera, are among the very oldest fossils known, occurring in Cambrian times. It is usually the case that such very ancient animals, which have come down to us practically unchanged from the remotest ages, are rare or at any rate local in their distribution, but *Nebalia* is a very striking instance to the contrary.

The other Crustacean type with which we have to deal, *Anaspides*, a peculiar fresh-water shrimp, is not known to be as ancient as *Nebalia*, though fossils practically identical with it are fairly common in Permian and Carboniferous formations. The surviving *Anaspides* and its allies (Fig. 10), unlike *Nebalia*, are extremely localized animals, occurring only in a few circumscribed localities in the fresh-waters of Tasmania and Southern Australia. The fossil forms, some of which are almost identical with the living *Anaspides*, inhabited the sea and are found in several localities in Carboniferous formations in England, Central Europe and North America. It was evidently a dominant form of shrimp at the time the Coal-measures were being deposited, but it is not met with in more recent formations, and we have no clue as to its history between the time that it peopled the Carboniferous seas and the present day when it survives only in a few tarns and streams on isolated

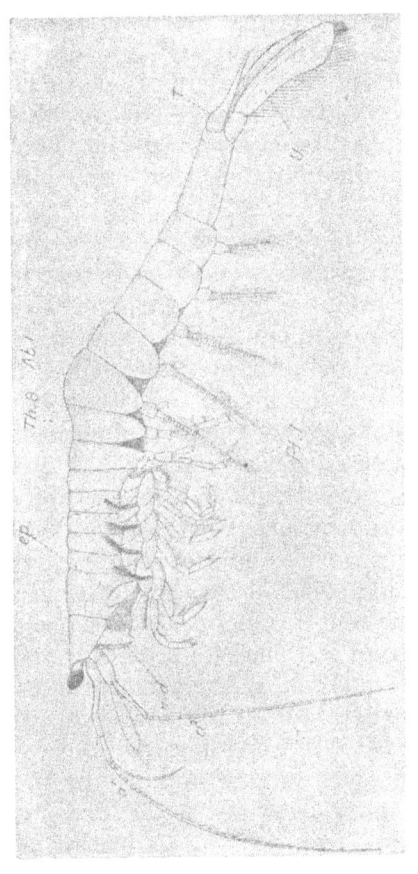

Fig. 10. *Paranaspides lacustris*, × 4. *a*¹, *a*², first and second antennae; *Ab.* 1, first abdominal segment; *ep.* epipodites or gills on the thoracic legs; *md.* mandible; *Pl.* 1, first pleopod; *T.* telson; *Th.* 8, eighth free thoracic segment; *U.* uropod, or sixth pleopod.

mountains in Tasmania. Nor have we any clue as to why this animal has disappeared from the sea and survives only in the fresh-waters of one small corner of the world, while the still more ancient *Nebalia*, as far as we know, has never penetrated into fresh-water nor suffered any diminution or restriction of its range in the sea.

The interest of *Anaspides*, besides its great antiquity and geographical isolation, is that it serves to connect the various groups of the higher Crustacea, the Malacostraca, with one another, and the study of its anatomy has done a great deal to show that the old classification of the higher Crustacea was erroneous, and to point out the true lines of cleavage. The older systematists drew a fundamental distinction between Malacostraca with a carapace and stalked eyes such as the Shrimps, Crayfish, and Crabs, and those without a carapace and with sessile-eyes, viz. the Isopods and Amphipods. *Anaspides* possesses stalked eyes but it is without a carapace, as its name implies; and in the rest of its characters it shows the same eclecticism, resembling in some features the more highly organized stalk-eyed forms and in others the sessile-eyed group. It is evident, therefore, that the structure of the eyes and the presence or absence of a carapace cannot be used as classificatory characters of the highest importance. This is borne out by the fact that certain marine shrimps, the Mysidae, which possess stalked eyes and a carapace, agree in their internal anatomy completely with the sessile-eyed, carapaceless Amphipods and Isopods. It would seem, therefore, that the latter owe their comparatively low type of organization to a secondary simplification of structure, and that the so-called higher Crustacea preserve to a larger extent the primitive character of the ancestral Malacostracan. The study of *Anaspides* supports this view, for whereas in many features it offers a starting-point for the divergence of the two groups, in other respects it possesses the organization of the more highly organized Malacostraca. Without going into details, it may be said that the general lesson to be learnt from *Anaspides* is that the evidence of its being of a primitive nature is found not in the absence of structures characteristic of the more modern and specialized groups, but in the possession of features which have been retained or elaborated in one of the higher groups and lost in the others. It cannot be too much insisted on that the history of evolution shows a balance-sheet in which, the better it is known, the more nearly are the losses seen to equalize the gains.

Besides the large and important groups of the Appendiculata which we have considered, there are several smaller assemblages of animals which have for long puzzled systematists but are probably degenerate or simplified offshoots from the Annelid stock. There is a small group of marine worms which at one time were thought to represent the primitive ancestral form of the Annelids, chiefly owing to their much greater simplicity of structure. This group includes the little worm *Polygordius* which does not possess a trace of limbs. It is now very generally recognized that *Polygordius* is descended from a marine Annelid which possessed limbs, and that it is one of a series of degenerate forms which have gradually lost one after another of the distinctive features which characterized their more highly developed Annelid ancestors. The final term in this series of degeneration is represented probably by the Rotifers or Wheel-animalcules, fresh-water organisms which afford a favourite object for microscopic study. These animals have lost all trace of limbs, of segmentation, and of a coelomic body-cavity, but we can trace a series from such a form as *Polygordius* through a number of still further simplified types (*Dinophilus, Histriobdella,* etc.), until we reach the Rotifers.

The Appendiculate phylum, from its vast extent and from its including so many hard-bodied animals likely to be preserved as fossils, illustrates, perhaps better than any other Invertebrate phylum, what degree of success may be expected from an application of the evolutionary theory, and to what limitations this procedure is liable. We have seen the important results which a single animal, *Peripatus*, has yielded, by indicating how the Arthropod type of body-cavity and vascular system has been derived from the Annelid. But *Peripatus* is of fundamental importance only in this respect. In dealing with the various classes of Arthropods, we found that *Peripatus* throws no light on the derivation of these classes from some common Arthropodan ancestor. Again, in dealing with these several classes we found that they existed in a fully differentiated condition in the oldest fossil-bearing rocks known to us, and that even existing families, if not genera, were present at the most remote geological periods and have suffered little or no important changes through all the incalculable ages that have supervened from that time to this.

We seem to be confronted by two apparently opposing bodies of fact; on the one hand, the immense antiquity and stability of living forms, on the other the evidence of a vast process of change and extinction. But at least we may have the satisfaction of knowing the reason of our perplexity, namely that we are trying to disentangle the plot of a drama of which we are permitted to be the spectators of only the closing scenes.

CHAPTER IV

EMBRYONIC AND LARVAL HISTORIES

THE doctrine that animals in the course of their development pass through or recapitulate the stages of their ancestral history, in other words that animals in their development climb up their own genealogical trees, has given rise to as much controversy as any biological theory. Von Baer, who is generally held responsible for this doctrine, did not enunciate it in this form; his statement was that in any two or more related animals the further back we go in their developmental history from the egg the more do they resemble one another. In this form it must be admitted that Von Baer's law holds good with very few exceptions, but the extension of this generalization to mean that the developmental stages represent actual adult animals, the ancestors of the particular species in question, is open to very grave objections and is indeed only partially true in a few cases. The fact that related animals on the whole differ less in their developmental stages from one another than in their adult form indicates that on the whole the process of structural change and evolution has affected the developmental and larval histories of animals less radically and variously than the adult structure, but the falsity of the extension of Von Baer's law depends upon the fact that the progressive change in an animal's structure does not necessarily take place in the adult stage alone, but may affect any or all of the processes of acquiring that stage. Even on purely theoretical grounds, we do not know if, quite apart from any active adaptive change in an animal's development, the mere alteration of the adult structure may not involve changes in the early structure and development.

However this may be, it cannot be denied that embryonic and larval histories often throw a quite unique light upon the relationship of adult animals, which could not be guessed in any other way, and this implies that the embryonic development of all the organs of an animal must be taken into account in estimating its relationship just as much as, if not more than, their adult structure.

As an example we may take the peculiar group of Cirripede Crustacea, the Rhizocephala, which live as parasites upon other Crustacea, chiefly crabs and hermit crabs[1]. The adult structure of these parasites is so highly degenerate that no one could guess that they were Crustacea at all; in fact the Neapolitan naturalist, Cavolini, who first discovered them, did

[1] The commonest of these parasites is *Sacculina*, found on the shore-crab (*Carcinus*) and on the spider-crab (*Inachus*).

not regard them as animals at all but as a tumour produced by the crab itself, and successive naturalists regarded them as lowly organized worms or molluscs. The adult structure consists of a bag attached beneath the crab's tail, in which the reproductive organs are situated and at maturity large quantities of eggs are produced; this bag communicates, at the point of attachment, with the interior of the host by means of a ramifying system of roots which penetrate to all parts of the crab's body and absorb nourishment from its blood.   There is nothing in this structure to suggest a Crustacean or indeed any particular group of animals.   But the eggs of the parasite hatch out as little free-swimming larvae possessing all the characteristics of the typical Crustacean larva, the Nauplius (Fig. 11 A), with its three pairs of append-ages, and not only is the Nauplius larva of the Rhizocephala a typical Crustacean Nauplius, but it shows certain structures, e.g. the lateral horns, typical of the Cirripede Nauplius, the larval form assumed by all Barnacles before they become attached to rocks, keels of ships, etc.

Now the Nauplius of the Rhizocephala, after swimming about in the sea for four days, undergoes a moult and becomes metamorphosed into a totally different larval form, the Cypris (Fig. 11 B), with a single pair of antennae and six biramous thoracic limbs.   This Cypris larva is absolutely characteristic

of the ordinary Cirripedes or Barnacles, being met
with nowhere else.   In the ordinary Barnacle the

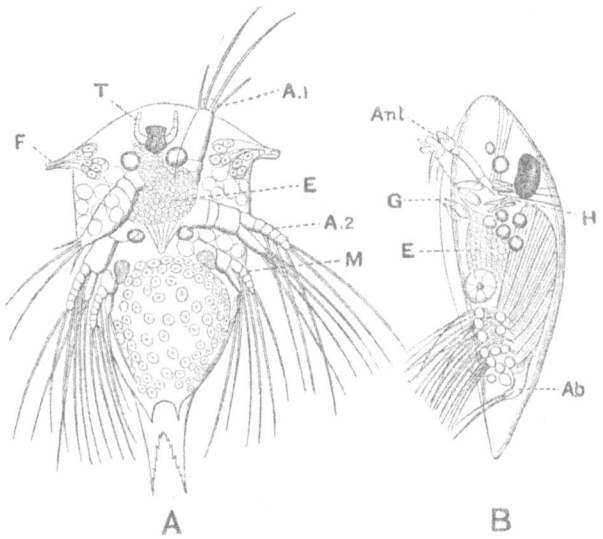

Fig. 11.   Development of *Sacculina neglecta*.   A. Nauplius stage,
× about 70.   B. Cypris stage, × about 70. A_1, A_2, 1st and 2nd
antennae of Nauplius; *Ab.* abdomen; *Ant.* antenna of Cypris;
E. undifferentiated cells; *F.* frontal horn; *G.* glands of Cypris;
H. tendon of Cypris; *M.* mandible; *T.* tentacles.

Cypris larva proceeds to fix itself by its antenna to a
rock or some floating object, and by a series of stages
it becomes metamorphosed into the adult sessile

Barnacle. In the Rhizocephala, on the other hand, the Cypris larva attaches itself by its antenna to a hair on the surface of a crab and passes into the crab a small group of cells which multiply rapidly in the blood-cavities of the host and finally produce the root-system and tumour-like body of the adult parasite which is thrust to the exterior (Fig. 12 A and B). The Cypris larva itself, which has thus infected the crab with a small portion of its body, does not play any further part but drops off dead. From our knowledge of the life-history of the Rhizocephala, therefore, we are in a position to affirm that these peculiar animals are really Barnacles which instead of fixing upon inanimate objects have taken to attaching themselves to living crabs and finally to infecting them and living as internal parasites within them.

Since we have made the acquaintance of the Nauplius larva, we may consider further the signifi-cance of this larval form in the Crustacea generally. We meet with it in nearly all the orders of the small, lower group of Crustacea, the Entomostraca; the majority of these animals hatch out from the egg as a Nauplius, which by the gradual addition of segments behind becomes transformed into the adult animal. In the majority of the higher Crustacea, the Mala-costraca, the development of the egg, before hatching, proceeds to a later stage, the Nauplius stage with three appendages being passed through within the egg membranes and not as a free-swimming larval form. But in a few of the Malacostraca, viz. in

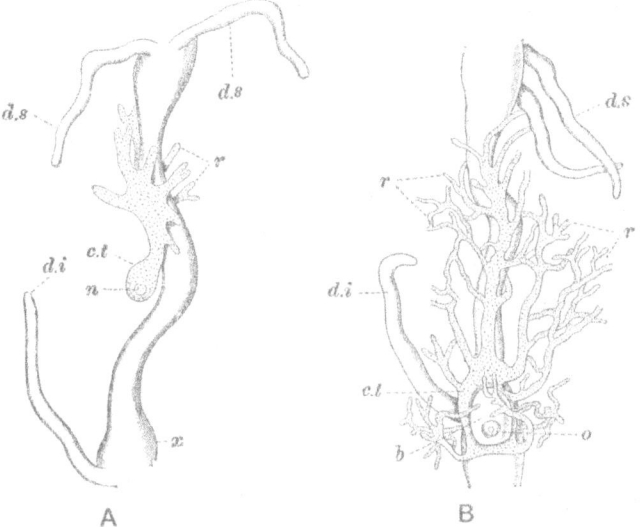

Fig. 12. A. The mid-gut of *Inachus mauritanicus* with a young *Sacculina* overlying it, × 2. *c.t.* "central tumour" of the parasite; *d.i.*, *d.s.*, inferior and superior diverticula of alimentary canal of host; *n.* "nucleus," or body-rudiment of *Sacculina*; *r.* its roots; *x.* definite position of the parasite. B. Later stage in the development of the "*Sacculina* interna," × 2. *b.* body of *Sacculina*; *c.t.* "central tumour"; *d.i.*, *d.s.*, inferior and superior diverticula of alimentary canal of host; *o.* opening of perivisceral cavity of *Sacculina*; *r.* its roots.

certain prawns, the animal hatches as a Nauplius, agreeing in all essential details of structure with the

Entomostracan Nauplius. This is a striking example of Von Baer's law and it tells us that the common ancestor of the Entomostraca and Malacostraca probably had a Nauplius larva. But it is not in the least justifiable to deduce, as some have attempted, that the Nauplius larva represents the adult ancestral form from which the Crustacea as a whole have sprung. The possession of only three segments and of the simple unfaceted eye, which are the really characteristic features of the Nauplius larva, do not carry us a degree nearer the common Arthropodan type from which the Crustacea, together with the Arachnids, Insects and Myriapods, must originally have diverged. Everything shows that such an ancestral type must have possessed numerous, probably similar, segments, and at least the beginnings of the compound faceted eyes, which occur under somewhat different forms in all the great Arthropodan classes.

There is another instructive Crustacean larva, the Zoaea (Fig. 13), characteristic of the Malacostraca, which, while admirably illustrating the truth of Von Baer's generalization, cannot be accepted as representing the type of the adult Malacostracan ancestor. The Zoaea larva is characterized by the possession of two pairs of antennae, mandibles, two pairs of maxillae and two biramous maxillipedes. The hinder part of the thorax is in a suppressed or retarded condition of development, but the abdomen is segmented and well formed and the last abdominal segment has its broad plate-like appendages precociously developed. Now although this larval form occurs with only minor

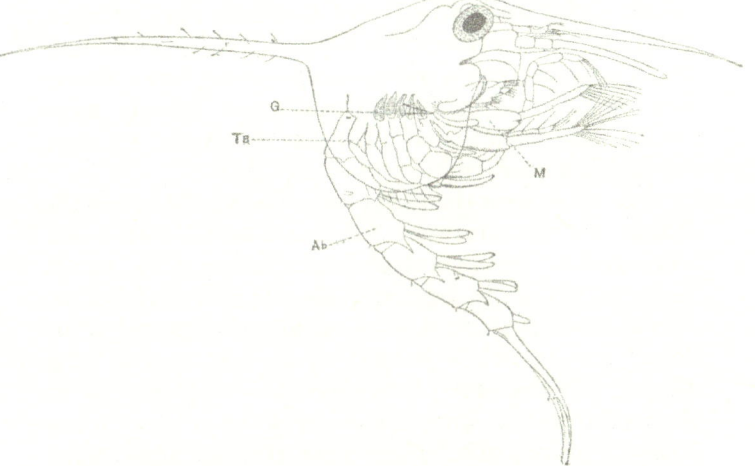

Fig. 13. Metazoaea, of *Corystes cassivelaunus* (×13). *Ab.* 3rd abdominal segment; *G.* gills; *M.* 1st maxillipede; *T* 8, last thoracic appendage. (After Gurney.)

variations in such widely different animals as prawns, hermit crabs and true crabs, and thus is in accordance with Von Baer's law, yet it cannot possibly be looked upon as representing the adult ancestral type of these animals. The retarded development of

S. A. K. 5

the thorax and the precocious development of the abdomen and tail fan cannot possibly be primitive features ; they are clearly specializations for the pelagic free-swimming larval life, providing the larva with suitable organs of locomotion while the development of the middle region of the body may proceed. It would be quite absurd to suppose that the ancèstral Malacostracan had a rudimentary thorax like a Zoaea, because the Zoaea in this respect is clearly more highly specialized and aberrant than any existing adult Malacostracan.

We have seen then that both the Nauplius and the Zoaea do not represent adult ancestral forms, but very ancient and ancestral larval forms. We cannot however press the argument too far and say that no larval form ever represents an adult ancestral form. Every case must be judged on its own merits after taking into account all the available evidence to be derived from the comparative study of related forms. Thus it seems indubitable that the so-called Mysis larva (Fig. 14), a stage assumed by many higher Malacostraca after the Zoaea, in which all the segments are fully formed and all the thoracic and abdominal segments bear biramous appendages, does very closely represent an adult ancestral form, the actual existence of which is nearly realized by the primitive *Anaspides* and Schizopod shrimps. Here we are dealing with a late larval form, just before the adult structure is attained, and it would appear not to have had time to have been profoundly modified in accordance with the larval existence.   It would seem, therefore, that we are more likely to meet with actual

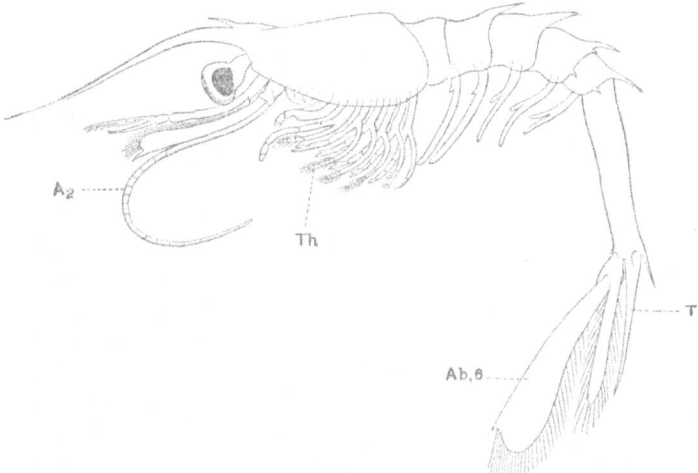

Fig. 14.   Mysis stage in the development of *Peneus*, sp. A_2, 2nd antenna; *Ab.* 6, 6th abdominal appendage; *T.* telson; *Th.* the biramous thoracic appendages.   (After Claus.)

adult ancestral types in the late larval or developmental stages of fairly closely related animals, which have not had time to be secondarily modified, than in the very early stages of more remotely related forms, although these, in accordance with Von Baer's law,

may show many points of resemblance which are lost
in the adults.

A larval form, yet more widely spread than either
of the Crustacean larvae mentioned, is the Trochosphere larva (Fig. 15), characteristic of Annelids of all
kinds and curiously enough of some Molluscs.   The
Trochosphere is a little spherical transparent organism which propels itself, generally near the surface of
the sea, by means of a band or circlet of cilia, which is
typically situated in front of the mouth, and is known
as the preoral ciliated band.   The sense organ is a
tuft of stiff hairs upon the upper pole of the organism,
known   as   the   apical   organ.   There is a curved
alimentary canal, the mouth being placed laterally
and the anus at the lower pole.   Internally the space
between the ectoderm and the gut is occupied by
small muscles, wandering cells, and often a pair of
excretory tubes or nephridia.   In addition to these
structures, which make up the active organization of
the larva, a solid mass of cells is present on each side
of the hind end of the body (Fig. 15 *c. mes.*), which are
destined to grow into paired segmented bands and to
form all those mesodermal structures of the adult
organism which are associated with the coelom, viz.
muscles, and reproductive organs.   Thus in the transition which occurs from the Trochosphere larva to the
adolescent   worm,   all   the   larval   organs   except   the
skin and the gut may be thrown away or absorbed

while the most important adult organs are formed anew from the paired bands of mesodermal cells, which become segmented into the hollow coelom sacs of the young worm.

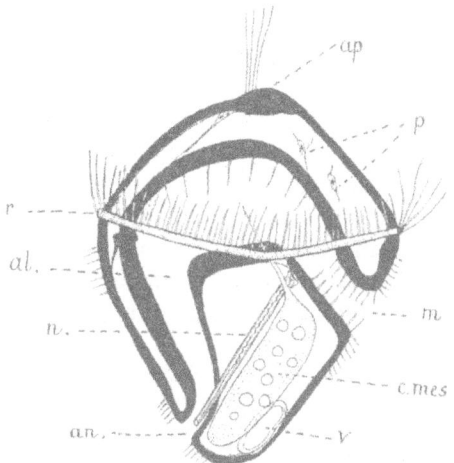

Fig. 15.   Diagram of Annelid Trochosphere, seen from the right side. *m.* mouth; *an.* anus; *al.* gut; *ap.* apical sense organ; *r.* preoral ciliated band; *n.* primitive kidney; *c.mes.* coelomic mesoblast; *v.* anal vesicle; *p.* mesenchyme cells lying in blastocoel. (Modified from C. Shearer, *Q. J. M. S.* vol. LVI.)

The Trochosphere, with modifications, occurs as a free-swimming larval form in a large number of Annelids, which are thus widely dispersed from the parent at an early stage of existence. Now it is of

especial interest that a similar larval form is assumed
by some Molluscs, which in the adult state are so
entirely different in their whole organization from
worms of any description.   This fact implies that the
Trochosphere is an exceedingly ancient type of larva,
which must have been possessed at some period by
the common ancestor of the Annelids and Molluscs,
and that this period is a very remote one is shown by
the fact that the adult Annelid and Mollusc at the
present time have nothing in common that would lead
us to suppose that they are in any way connected.
There is however nothing to support the view that the
Trochosphere in any way represents the common
ancestor of the Annelids and Molluscs, any more than
that the Nauplius represents the ancestor of the
Crustacea.

The Echinoderms or Star-fishes, Sea Urchins, etc.,
which were mentioned in the first chapter as being
among the few animals possessing a radial as opposed
to a bilateral symmetry, pass through a free-swimming
larval phase known as the Auricularia (Fig. 16), which
differs entirely from the Trochosphere.   The Auricularia like the Trochosphere is a small transparent
marine organism, with a curved alimentary canal and
a ciliated band, but this ciliated band is post-oral in
position, encircling the body with a sinuous girdle
which lies between the mouth and anus.   Internally
the Auricularia is also distinguished by the possession

of typically three paired coelomic pouches (Fig. 16 *a.c., m.c.* and *p.c.*), which are derived as paired hollow outgrowths from the gut.   It will be noticed that this mode of origin of the coelom is very different from that which occurs in the Trochosphere larva, where

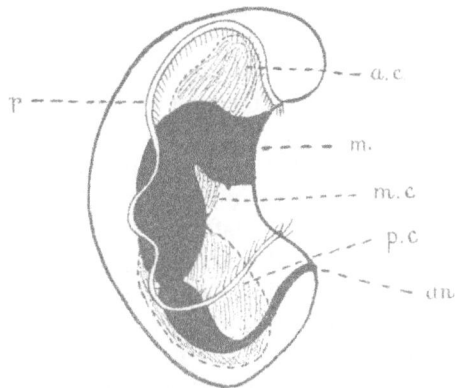

Fig. 16.   Diagram of Auricularia larva of an Echinoderm, seen from the right side.   *m.* mouth; *an.* anus; *r.* post-oral ciliated band; *a.c.* anterior left coelom sac which will be converted into the axial sinus; *m.c.* middle left coelom sac which will be converted into the hydrocoel; *p.c.* posterior left coelom sac which will form gonocoel or the reproductive organs.   The corresponding coelom sacs of the right side are not shown.

the coelomic mesoblast forms two paired solid bands of cells which become subsequently segmented and hollowed out to form the segmented coelom sacs of the adult.   The fate of the coelom pouches of the Auricularia is also very different from that of the coelom in the Trochosphere, as the two anterior pairs of pouches are transformed into a system of fluid-containing vessels which function as a kind of circulatory system. The posterior pair of pouches are, however, transformed into the reproductive organs, and in this respect they agree with the coelom of all coelomate animals, in which the reproductive organs are invariably formed from the coelom.

Another point of difference between the Trochosphere and the Auricularia is found in the constant absence of an excretory nephridium in the latter.

The Auricularia is a bilaterally symmetrical organism, but in its transformation into the adult form of Echinoderm it becomes peculiarly twisted and asymmetrical at first, and then a new type of symmetry, the radial, is imposed upon it, producing the peculiar forms of the Star-fishes and Urchins and Sea-lilies with which everyone is familiar.

Now it is a most noteworthy fact that an animal differing entirely in the adult state from the Echinoderms, viz. *Balanoglossus* (a worm-like burrowing creature which is regarded as being most nearly related to the Vertebrata, owing to its possession of such characteristic Vertebrate organs as gill-slits and of certain other structures), possesses a free-swimming transparent larval form, the Tornaria, which is practically identical in all its features with the Auricularian larva of Echinoderms. There is absolutely nothing in the adult structure of *Balanoglossus* that could possibly lead anyone to connect it with the adult Echinoderm, and yet the larval forms of these two totally different kinds of animals are for all purposes identical. We cannot avoid the conclusion that this identity proves that *Balanoglossus* and the Echinoderms at some remote epoch possessed a common ancestor with this larval form, which has been preserved during the immense period in which the divergent evolution of the two adult types has proceeded. The conclusion surely involves a very great antiquity for this larval type, which thus is seen to be a link with the past of a very persistent kind.

We have touched so far only upon certain ancient larval forms, but we might have illustrated the antiquity of many developmental phases by reference to the embryonic stages of many animals, passed in a quiescent state within the egg membranes. Space will only permit us to refer to one such instance, namely the persistent development of gill-like structures in the embryos of all Vertebrate animals, even in those Vertebrate animals, such as birds, reptiles and mammals, which never spend any part of their existence in the water breathing by means of gills.

The whole structure of the head in these animals is fundamentally affected by the fact that in the young embryo at least three and usually more gill-clefts are formed with their associated supporting structures, blood-vessels, nerves and muscles, in which all the principal parts can be homologized with corresponding parts in the gills of fishes. The subsequent development of the head in this region takes place by a gradual alteration of the gill-structures into various organs associated with the ear and throat, which finally in the adult subserve totally different functions from those which they were originally fashioned to perform. It may seem strange that it has sometimes been argued that this does not imply that the ancestors of these land vertebrates were fish-like creatures and breathed by functional gills. It is true that if we were desirous of exhibiting an excess of caution we might argue that it is only safe to conclude that the ancestor of the land vertebrates, at some stage in its life, viz. a larval stage, breathed by functional gills, and that the adult ancestor never did so. But our excessive caution in this instance would be singularly misplaced; as it would involve one of two equally unlikely and gratuitous suppositions, either that the class of Fishes has no connection at all with the higher vertebrates, or else that the Fishes were originally land animals with a fish-like larval stage, which have retrograded in development and lost all the higher terrestrial phases of their organization. Needless to say there is no shadow of evidence for either of these suppositions, and though morphological science may not carry with it the conviction of exactitude, it is not necessary on that account to make it palpably absurd.

CHAPTER V

THE ANCESTRY OF THE VERTEBRATES

BEFORE it was clearly recognized that the Invertebrate phyla were separated from one another by fundamental differences, it was customary to regard the distinction between Invertebrate and Vertebrate animals as constituting the great cleavage or dichotomy of the animal kingdom. At the present time, although we may fully admit that among Invertebrate animals cleavages of equal magnitude may exist, yet the fact that among the Vertebrata are included all the highest developments of animal life culminating in man himself, has invested and still invests the question of the origin of the Vertebrate stock with peculiar interest. From what existing phylum of Invertebrata, if from any, did the Vertebrate type take origin? That is the question which in the past and present has been debated with vigorous and sometimes almost ecstatic fervour. The true Vertebrates include all those animals which possess a dorsal tubular central nervous system, a segmented backbone or vertebral column enclosing the nerve-cord, a closed vascular system containing blood carrying a special type of corpuscle charged with respiratory haemoglobin, a ventrally situated contractile heart, and a spacious coelomic body-cavity. The pharynx is also pierced or nearly pierced at some period of the animal's existence by a series of paired gill-slits, which may thus put the cavity of the pharynx into communication with the exterior. Such animals are Fishes, Amphibia, Reptiles, Birds and Mammals.

Now besides these animals there are others, which, while not displaying all the above-mentioned characteristics, yet betray their affinity to the Vertebrate class by the possession of some of them.

We designate these animals in common with the Vertebrata as Chordata, and they together with the true Vertebrata or backboned animals constitute the Phylum Chordata. It is among these animals, as representing a lower grade of organization than the Vertebrata, that we might expect to find some indication of the origin of the Chordata from an Invertebrate stock; we will therefore consider them in some detail.

One class of these lower Chordates, the Ascidians, we may dismiss with a few words, because it will be evident that they owe their simplicity of structure chiefly to degeneration incident on a special mode of life.

The Ascidians, known popularly as Sea-squirts by

those who are familiar with common marine animals, live for the most part a sedentary life fixed to objects such as rocks or the piles of a pier, and they gain their subsistence by passing a current of water, charged with living organisms, through their pharynx which is pierced by innumerable gill-slits. Like so many fixed animals, their nervous system is degenerate, consisting of a single ganglion placed dorsally to the mouth. This is the condition in the adult state of the majority of Ascidians, but in the young stages, after hatching from the egg, they lead an active free-swimming existence, when they appear with all the characters of a small tadpole, pre-eminent among these characters being the possession of a supporting rod, the notochord, which forms the principal skeletal or supporting tissue, especially for the tail which acts as the chief locomotive organ. This notochord, which occurs in the young stages of all Chordate animals, being the forerunner of the vertebral column or backbone of the Vertebrates, enables us to place the Ascidians in the Chordate phylum with absolute certainty. When the Ascidian tadpole becomes fixed and sessile, preparatory to assuming the adult state, the notochord is absorbed and disappears.

The next class of Chordate animals is occupied by a single form, *Amphioxus* or the Lancelet, and this organism is of the most fundamental importance, since it undoubtedly represents in many of its features a primitive or ancestral form of Chordate. The Lancelet (Fig. 17) is a little cigar-shaped marine animal, attaining an inch or two in length ; it swims with great activity and when alarmed burrows with extraordinary rapidity into fine sand. It is found in small numbers round the English coasts, but with much greater frequency in the Mediterranean and tropical seas. It possesses the essential organization of a very simple fish, but the simplicity of its structure, especially as regards sense organs, brain, skeleton and vascular system is carried to such an extent, that the actual resemblance to a fish is extremely shadowy. The skeleton is represented solely by the long notochord which stretches, as a flexible gelatinous rod from the tip of the snout, along the back, to the tip of the tail. There are no skull, or jaws, or paired limbs, though there is a fin running along the back and expanded at the tail. Thus as compared with a fish the skeleton is in a rudimentary state, being represented solely by the notochord. This organ, which is always present in the embryo Vertebrate, is invariably replaced in the adult Vertebrate, either partially or entirely, by the segmented vertebral column or spine, which grows over and encloses the nerve tube. The nervous system is in the form of a hollow tube running along the back, just above the notochord. In the head it gives off two pairs of

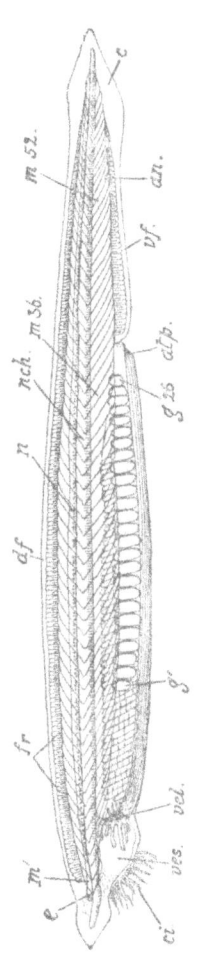

Fig. 17. An adult specimen of *Amphioxus lanceolatus*, seen from the left side as a transparent object. *an.* anus; *atp.* atriopore; *c.* caudal fin; *ci.* buccal cirrhi; *df.* dorsal fin; *e.* eyespot; *fr.* fin-rays; *g¹–g²⁶*, the twenty-six pairs of gonadial pouches; *m¹*, the first, *m³⁶*, the thirty-sixth, *m⁵²*, the fifty-second myotomes; *n.* neural tube; *nch.* notochord; *vel.* velum, in front of it are the finger-like processes of the wheel organ; *ves.* vestibule; *vf.* ventral fin. (From G. C. Bourne.)

nerves with single roots which are purely sensory in function and supply the skin in the head region. All the succeeding nerves have two roots, a dorsal and a ventral, thus corresponding to the spinal nerves of the Vertebrates. It is thus seen that the central nervous system of the Lancelet is in a much simpler condition than that of the lowest Vertebrates, the Fishes. In them the front end of the central nervous system is swollen to form a complicated brain from which the first ten nerves arise as cranial nerves, specialized in connection with the elaborate sense organs present on the Vertebrate head. In the Lancelet the sense organs consist of a very simple eye and a small pit, doubtfully an olfactory organ, both of which are directly applied to the substance of the brain. The brain itself is simply the un-differentiated front end of the nerve-cord, and the only nerves corresponding to the cranial nerves of the Vertebrates are the two most anterior pairs. The alimentary canal is again simpler in arrangement than that of the Vertebrate; it is not coiled, and it has only one digestive gland in connection with it. The pharynx is pierced by very numerous gill-slits (Fig. 18) enabling water to be passed out of it to the exterior, and in this respect it has a schematic resemblance to a Fish, though the gills and gill-bars do not resemble those of a fish in any detail.

The vascular system, in which a colourless blood circulates, is very simple ; there is no differentiated heart, all the larger vessels being feebly pulsatile ; and we can only say that it resembles the Vertebrate vascular system in being closed off from the coelom and in the fact that the blood passes forwards in the ventral vessel and backwards in the dorsal.

We now pass to the condition of the body-cavity and musculature. The first thing that strikes one on looking at a Lancelet is that the body is cut up transversely into a series of narrow zig-zag strips (Fig. 17). If these strips were straight instead of zig-zag, we should say that the body was formed of a series of rings, as in the Earthworm, and we should suspect that the animal was metamerically segmented. Now if we cursorily examine a fish or any other Vertebrate, we might not immediately perceive, what is in fact the case, that all Vertebrates, too, are essentially metamerically segmented animals. Yet, on reflection, the fact is plain. The vertebral column or spine of the Vertebrate is formed of a series of repeated homologous parts, the vertebrae. Corresponding with these vertebrae there issue from the central nerve-tube a series of repeated homologous spinal nerves, just as from the segmented nerve ganglia of the Earthworm, or any other Appendiculate animal, there arise nerves which pass to the corresponding segments of the body. But, as in the

Appendiculata, the body muscles of the Vertebrate are also metamerically segmented, being derived in the embryo from the walls of the coelom which in the Vertebrate, as in the Earthworm, is cut up into repeated, metamerically segmented, blocks. The only difference between the metameric segmentation of the Vertebrate and of the Annelid is that in the former only the dorsally situated parts of the coelom, the myotomes, are involved in the segmentation, the perivisceral coelomic cavity being unsegmented, whereas in the latter the coelom is completely divided, both dorsally and ventrally, by a series of septa which partition the perivisceral coelom into metamerically repeated compartments. This is, however, a small difference and need not obscure the central fact that Annelids and Vertebrates, including *Amphioxus*, resemble one another fundamentally in the type of their metameric segmentation.

There is still another feature in which *Amphioxus* betrays a similarity to the Annelid, namely in its excretory organs (Fig. 18 *nph.*). These organs in *Amphioxus* are a series of metamerically repeated tubes which end in closed excretory cells, known as solenocytes, bearing an extraordinarily accurate resemblance to the excretory cells which terminate the segmental nephridia of many marine Annelids. There can hardly be any doubt that these excretory tubes in *Amphioxus* are true nephridia of ectodermal origin,

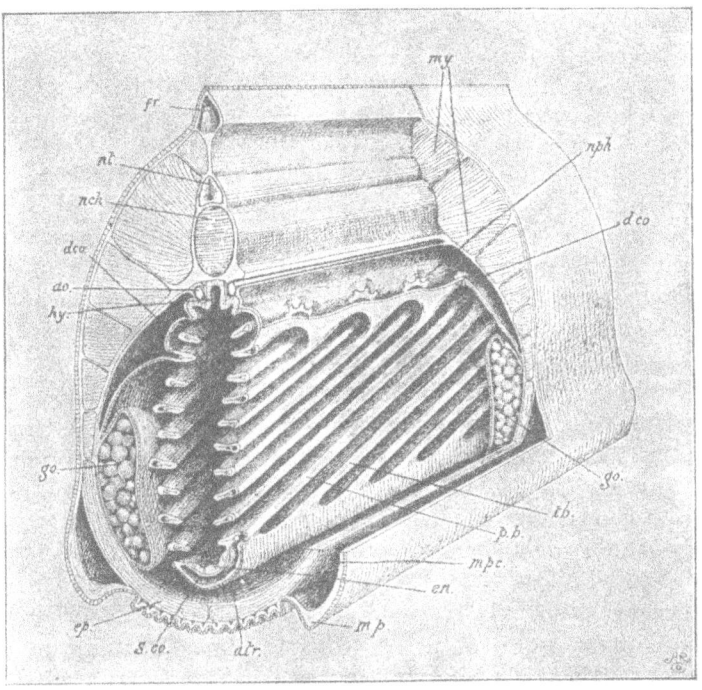

Fig. 18. Diagram illustrating the anatomy of the pharyngeal region
of *Amphioxus*. A segment has been cut out of the pharyngeal
region, and a portion of the right wall removed to show the atrial
cavity, the pharynx and its gill-slits, and the dorsal coelom;
ao. dorsal aorta; *atr.* atrium; *d.co.* dorsal coelom; *en.* endostyle;
ep. epipleur; *fr.* fin-ray; *go.* gonads; *hy.* hyperbranchial groove;
mp. metapleur; *mpc.* metapleural canal; *my.* myotomes; *nch.*
notochord; *nph.* nephridium; *nt.* neural tube; *p.b.* primary gill-
bar; *tb.* tongue-bar; *S.co.* sub-endostylar coelom. (From G. C.
Bourne.)

6—2

and that they are homologous to the Annelid nephridia.

In the true Vertebrates, these nephridia have apparently disappeared without leaving a trace of their existence, and they have been replaced by a segmental series of coelomoducts, little tubes which arise as outgrowths from the dorsal peritoneum, and some of which become conglomerated to form the adult kidney, while others take up an intimate relation to the reproductive organs and function as genital ducts. A similar interchange of function is of frequent occurrence in the Annelids, in which coelomoducts may be either genital or excretory in nature.

Our short conspectus of the structure of *Amphioxus* has shown us that whilst this animal is closely related to the higher Vertebrates, yet in other respects it strongly points to Annelidan affinities, and even apart from *Amphioxus* we might be tempted to associate the Annelids and Vertebrates in a common origin from the profound similarity which exists between them in the matter of their metameric segmentation.

We are led therefore on evidence of considerable weight to adopt what is known as the Annelid theory of Vertebrate descent. There is however a grave, and to some thinkers an insuperable, objection to this theory. *It is that the relative position of the gut and of the central nervous system in the Annelid and in the Vertebrate are exactly and diametrically reversed.* In the Annelid the brain is dorsal to the alimentary canal while the nerve-cord is ventral. In the Vertebrate if we regard the brain as dorsal, then the central nervous cord runs along the back and is also dorsal to the alimentary canal. A diagram will make this clear (see Fig. 19 A, B and C).

It is possible to overcome this difficulty, but the process of overcoming it involves an hypothesis which to many appears to overstep the bounds of temerity. In order really to homologize the Annelid part by part with the Vertebrate, we must suppose that the dorsum of the Vertebrate corresponds to the ventrum of the Annelid, that the brain of the Vertebrate does not correspond to the brain of the Annelid but to its suboesophageal ganglion, and that the mouth of the Annelid has migrated in the Vertebrate from its original ventral to a terminal and finally a dorsal position. This is made clear in the diagrams presented in Fig. 19.

Although it must be admitted that the hypothesis is a bold one, it cannot be charged with any inherent absurdity from a morphological standpoint. It may seem a revolutionary, almost impious, idea that what we have been accustomed to look upon as the back and the front of ourselves and other Vertebrates, have all along been diametrically the opposite ; that

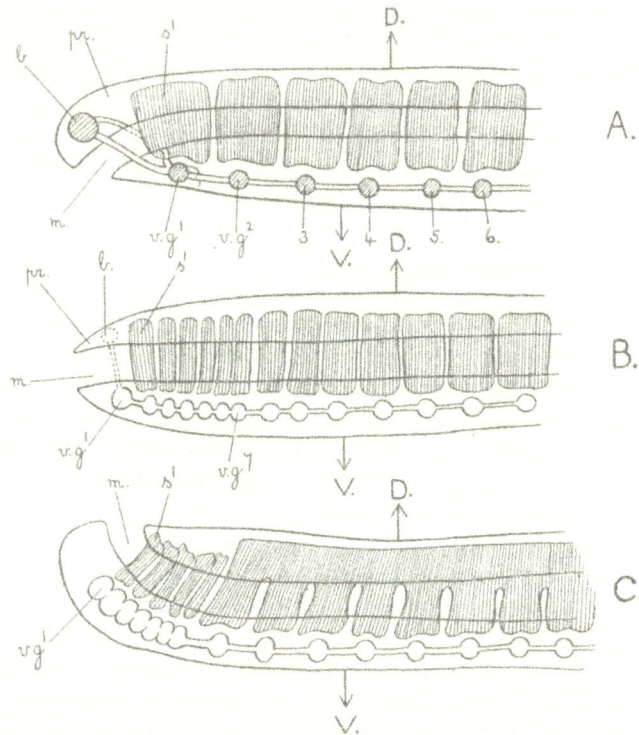

Fig. 19. Diagrams representing the derivation of the Vertebrate (C) from the Annelid (A) through the intermediate stage (B). In A there is a prostomium *pr.* containing a dorsally situated brain *b.*, connected by circum-oesophageal commissures passing round the mouth *m.* with the ventral chain of ganglia $v.g.^1$, $v.g.^2$, etc. The mesodermal somites $s.^1$ are not concentrated in head region.

In B, the mouth *m.* has shifted to a terminal position, the prostomium is reduced and the brain has disappeared with its commissures. A new brain is forming ventrally by the concentration of the ventral ganglia $v.g.^1$ to $v.g.^7$, and the somites are becoming crushed together in the head region. In C, the Vertebrate condition is attained by the mouth shifting on to the dorsal surface (which becomes the new ventral surface); the brain is formed by the further concentration of the ventral ganglia; the somites are still more crowded in the head and the dorsal parts of the posterior somites fuse together to form the continuous perivisceral coelom. Thus the so-called dorsum of the Vertebrate is shown in reality to correspond to the ventrum of the Annelid and *vice versâ*, if this view is accepted.

when, for instance, we imagined that we were exhibiting a bold front to our enemies or turning our back in reprobation, we were really acting in a manner unworthy of our ancestry. A kind of moral value has come to be associated with the ideas of back and front, and we allow this moral value to influence our morphological conceptions. But if we put ourselves in the position of a worm which for some reason is forced to adopt a more erect attitude, or at any rate to forsake its creeping habit, we should be hard put to it to know which surface to present to a hostile and not incurious world.

The complete loss of the original brain and prostomium, which is involved in the Annelid theory of Vertebrate descent, and its replacement by the ventrally placed suboesophageal ganglion, is not such a radical change as might be supposed. In the Annelids there is a decided tendency for the segments behind the mouth, which are innervated by the suboesophageal ganglion, to shift forwards and to take on the sensory functions of the prostomium. We are therefore far from dismissing as absurd the Annelid theory of Vertebrate descent, despite the morphological difficulty of the apparent change of the dorsal into the ventral surface thereby involved.

There is however another, and possibly more serious, difficulty in the way of its acceptance.

The peculiar worm-like burrowing animal known as *Balanoglossus* has for long been associated with the Chordata on account of a number of structures which it possesses and which point to a fairly close relationship with *Amphioxus* and so with the Vertebrates. In the first place this animal possesses gill-slits of an almost identical structure with those of *Amphioxus*; an organ is also present which suggests a notochord, and the arrangement of the coelom has been compared with that found in *Amphioxus*. Traces of a dorsal nervous system, of tubular structure as in Vertebrates, are present, and the arrangement of the gill-slits suggests the appearance of metameric segmentation.

Now this animal, as we have seen in the last chapter, pp. 72, 73, passes through a larval stage, the Tornaria, which in all its features recalls the exact structure of the Echinoderm larva, the Auricularia. If we admit the close relationship of *Balanoglossus* to the Chordates, the occurrence of this larva is a severe blow to the Annelid theory of Chordate descent, because the Tornaria larva in all its characters is the very opposite to the Trochosphere larva of the Annelids (see pp. 69, 71).

It is in fact impossible to associate the Echinoderms with the Annelids, in the ancestry of the Chordates, as they really possess nothing in common. If we persist in upholding the Annelid theory, there is nothing for it but to deny the relationship of *Balanoglossus* to the Chordates, and this is a measure which few would be confident enough to take.

This difficulty is in reality the great crux that besets the problem of the origin of the Chordata from any existing Invertebrate phylum. It is not from lack of resemblances between the Vertebrata and Invertebrata that we are at a loss ; there is much to be said for the Annelid theory, and much also for the inclusion of *Balanoglossus* in the Chordata, but the acceptance of the one excludes the other. Further than this unsatisfactory conclusion we are not prepared to go, and we must consign the question of the origin of the Chordata to the category of unsolved morphological problems.

There are certain other animals of obscure relationships which are held to belong to the same stock as the Echinoderms and to *Balanoglossus* and the Chordates.

The deep-sea organisms *Cephalodiscus* and *Rhabdopleura* are most nearly related to *Balanoglossus*. Of a much more obscure nature is the Phylum Brachiopoda, but mention may be made of them here as they afford perhaps the best example of a persistent type to be met with in the whole animal kingdom. The Brachiopods have the body enclosed in a bivalve shell, very closely resembling that of a Lamellibranch Mollusc (e.g. a Cockle), with which group they were originally classified, but their internal anatomy is not built on the Molluscan type at all, so that the resemblance of the shell in the two classes of animals is certainly due to convergence and not to a common ancestry. It is one of those instances of a close resemblance in superficial structure which the morphologist has continually to be on his guard against. The internal anatomy of the Brachiopod is not very simple ; it is a coelomate animal and it exhibits the same tripartite division of the coelom which we met with in the Echinoderm larvae and in *Balanoglossus*. It is held on this and other grounds that the Brachiopods are obscurely related to the stock which has given rise to Echinoderms, *Balanoglossus* and the Chordates, and that they have nothing to do in origin with the Appendiculata. There is a form of Brachiopod called *Lingula* which burrows in the sand between tide-marks or at a shallow depth and is found in certain localities off the coasts of many tropical and temperate countries.   On certain parts of the coast of Northern Canada the cliffs are formed of rocks belonging to a stratum lying at the base of the Cambrian and the very oldest in which fossils of any description are to be found.   In the cliffs it is possible to chip out very perfect fossil shells of the *Lingula* which lived on this coast in the pre-Cambrian period; from the sand at the foot of the cliffs we can dig out living specimens of *Lingula*, and if we should examine the fossil shells and those of the living animal with the minutest care we would not be able to detect the smallest difference, down to the finest striae, between them.   We have here revealed to us a fact which should never be lost sight of in considering the evolution of animal life, a fact which proves how short a period of the whole time in which life has existed on the planet is represented by the fossil-bearing rocks, and how stable and persistent certain types of animals have remained in the midst of the sweeping changes proceeding around them.

CHAPTER VI

THE ORIGIN OF THE LAND VERTEBRATES

No one disputes the truth of the fact that the land Vertebrates, which occupy the most exalted position in the scale of living organisms, must have originated from marine animals possessing the essential characteristics of fishes. All the evidence of palaeontology, of comparative anatomy and embryology, converges towards this cardinal tenet. But between the fishes and the lowest forms of land Vertebrates, the Amphibia, there is a serious gap which palaeontology does nothing, and comparative anatomy not everything, to bridge.

In the passage from the water to the land that must have occurred, two main systems of organs have been principally affected, the limbs, owing to the change of medium in which progression had to take place, and the respiratory organs, which had to change from the process of obtaining oxygen from the water to that of obtaining it directly from the atmosphere. As a matter of fact comparative anatomy enables us to trace the transition in the respiratory organs with far greater certitude than in the case of the limbs.

Besides the tail and the less important median fins upon the back and belly, fishes possess paired front and hind limbs, the pectoral and pelvic fins, which are certainly the forerunners of the arms and legs of terrestrial Vertebrates.   These paired fins of the fish do not however resemble in any detail the limbs of the land animals.   They are flattened structures with either a short and broad, or long and segmented axis to which are attached numerous rays serving to stiffen the membrane of the fin.   They are of course used as paddles to propel the animal through the water or, where the tail and body give the chief motive force, to serve as guiding rudders. There are several cases of fishes which have accessory locomotory structures to enable them to jump on dry land, to creep over rocks, and even to climb trees. The little goggle-eyed fish, *Periophthalmus*, jumps about on coral reefs out of the water, by means of flicking its pectoral fins ; some Gurnards (*Trigla*) can creep about on the sand by means of certain finger-like rays on their pectoral fins ; the climbing perch of Ceylon, *Anabas scandens*, can progress on dry land and ascend trees by helping itself along with spines growing out from its gill covers ; but none of these structures have any more bearing on the acquisition of the land Vertebrates' limbs, than the fins of a flying-fish have on the derivation of the wings of a bird.

The curious thing about the true land Vertebrate limb is that it appears apparently for the first time in the Amphibia in its complete and fully formed condition, consisting in each limb of a single bone, the humerus or femur, jointed to the pectoral or pelvic girdle respectively, followed by two parallel bones, the radius and ulna in the fore limb, the tibia and fibula in the hind limb, and then come a number of small bones forming the wrist and ankle, and finally five digits, the fingers and toes. That is the uniform structure of the pentadactyle limb of the terrestrial Vertebrate, which appears suddenly, to all appearances from nowhere, and is preserved bone for bone with extraordinary uniformity in all the land Vertebrates, including man.

Of course we must believe that this pentadactyle limb had an origin, and that it was somehow derived from the fishes' fin, but there is no shadow of a suggestion, either in living forms or in the geological record, of the steps by which this fundamental acquisition was arrived at.

Let us now turn to the consideration of the origin of the breathing organs of the land Vertebrates. The typical fish obtains its oxygen from the gas dissolved in the water, a fresh stream of water being continuously passed over the gills, the walls of which are very thin and vascular, so that the blood circulates in them in close proximity to the oxygenized water. The gills consist of a series of pouches opening inwardly into the throat and outwardly to the exterior; the pouches have folded walls, the folds being produced into fine lamellae or filaments which are charged with blood, and they are supported by cartilaginous bars, or gill-bars, which give rigidity to the gill walls and support the branchial filaments. The number of gills varies in the different kinds of fishes; they are more numerous in the Selachians (Sharks, Dogfishes and Rays), in which there may be as many as seven gill-pouches, while in the more modern Bony-fishes or Teleosts there are only four. In the Selachians the gill-pouches open separately by slits to the exterior and the first gill-cleft is modified to form a special tube, the spiracle, which opens in front of the first gill arch or hyoid arch, immediately behind the eye. This gill-cleft carries only rudimentary gills, and it serves under certain conditions for the intake of water into the pharynx. In the other fishes the spiracular gill-cleft does not open to the exterior and is quite rudimentary, while in the land Vertebrates it becomes converted into the cavity of the middle ear and its opening into the throat remains as the Eustachian passage. This remarkable transformation will be considered later.

The gills of the fish are supplied with blood by a system of afferent arteries, branching off from

the ventral aorta, which convey venous, i.e. de-
oxygenated, blood from the heart. The blood is
oxygenated in the capillaries of the gills, and is then
brought back by a series of efferent branchial
arteries which join to form the dorsal aorta, and so
the arterial or oxygenated blood is distributed to the
various organs of the body.

The ordinary fish is entirely adapted to breathing
in the water, so that exposure to the air in a very
short time leads to death by asphyxiation.

But just as some fishes can progress upon the dry
land, so there are fishes which can withstand suffo-
cation during a prolonged absence from their natural
element. The fresh-water Siluroid fishes, which occur
in the tropics, *Clarias* and *Saccobranchus,* have
special chambers leading from their gill-pouches
which are adapted for aerial breathing, and it seems
that these fishes, even when living in the water, take
periodic gulps of air without which they may become
asphyxiated in the stagnant muddy pools in which
they dwell. Many examples might be adduced of
fish having their gill chambers modified for aerial
respiration, but of more profound interest are those
fishes which make use of a separate organ, the air-
bladder, for respiration, since this organ is undoubtedly the homologue or forerunner of the land
Vertebrate's lung. The air-bladder does not occur in
the Selachians, but in bony fishes of all sorts it is a

characteristic organ, though it has disappeared in an erratic and unaccountable way in many genera.   The air-bladder of the Teleostean fish, while exhibiting considerable variations in its detailed structure and anatomical relations, is essentially a hollow thin-walled sac produced as a large pouch from the oesophagus, and often communicating with the throat by means of a duct, the pneumatic duct.   The bladder itself is in all cases save one disposed as a spacious air-containing bag dorsally to the oesophagus and viscera, being covered over by the dorsal peritoneum. In *Polypterus*, the Bichir of the Nile and West African rivers, a peculiar and primitive fish, both bladder and duct are situated ventrally to the oeso-phagus, but in all other fishes the bladder has travelled on to the dorsal surface, though the duct frequently remains in its original ventral situation, travelling dorsally round the oesophagus to the bladder (see Fig. 20 B).

In modern Teleosts the air-bladder is aerostatic in function, the lining secreting a mixture of oxygen and nitrogen gas, the nitrogen being usually greatly in excess.   The function of the bladder in these forms consists in the power it gives the organism of counter-acting the effects of varying pressure at different depths of water.   The arterial blood supply to the air-bladder in the ordinary Teleost or Bony-fish (Fig. 20 D) is given off from the coeliac artery, and the blood is returned to the general circulation by two small veins passing to the posterior cardinal and the portal veins. In these fishes the air-bladder has no respiratory function.

In the primitive Teleosts, such as *Polypterus*, the blood supply is different, the arterial supply coming from the sixth or posterior pair of branchial arteries, while the veins open into the ductus Cuvieri which passes directly to the heart. In these fishes the air-bladder probably exercises some respiratory function, but the complete transformation of the air-bladder into the lungs is effected in the curious fishes known as the Dipnoi. In the Dipnoi (Fig. 20 B), or Lung-fishes, the air-bladder, which lies dorsally to the oesophagus and opens into it ventrally by a curved pneumatic duct, is a median or bilobed sac with vascular and honeycombed walls like a lung, and it is supplied by a pair of pulmonary arteries which arise from the sixth branchial arteries, and a pair of veins which empty directly into the heart. The position of the air-bladder and its duct in these fishes is exactly the same as that of the lungs and trachea in a true land Vertebrate ; the minute structure of its walls is the same, and its blood supply is also essentially the same, since the pulmonary arteries in the land Vertebrate always arise in the embryo from the sixth branchial arches, and the pulmonary veins pass back directly to the heart. There cannot be any doubt therefore that the air-bladder of the Dipnoi is the homologous fore-runner of the land Vertebrate's lung, and that its

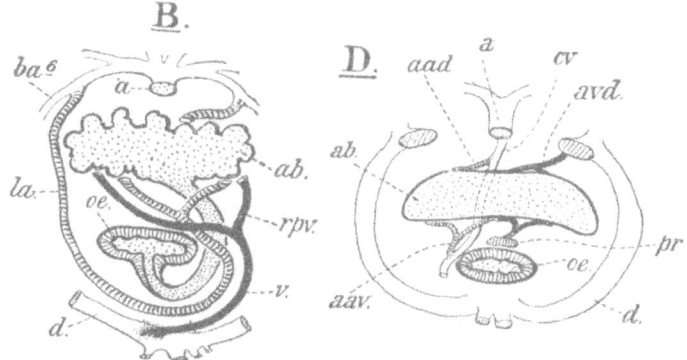

Fig. 20, B and D.   Diagrams illustrating the blood supply of the air-bladder in B, *Ceratodus*, and D, a Teleost.   The blood-vessels are seen from behind and are cut short in transverse section. *a*. dorsal aorta; *aad*. anterior dorsal artery from the coeliac; *aav*. anterior ventral artery; *ab*. air-bladder; *avd*. anterior dorsal vein to the cardinal; *ba⁶*, 4th aortic arch; *cv*. coeliac artery; *d*. ductus Cuvieri; *la*. left pulmonary artery; *oe*. oesophagus; *pr*. portal vein; *rpv*. right pulmonary vein; *v*. left pulmonary vein. (After E. S. Goodrich, Lankester's *Treatise*.)

function is already the same, namely for the supply of atmospheric oxygen to the blood.

The habits and nature of the Dipnoi are of such interest that we may describe them in some detail. There are three species of Dipnoi or Lung-fishes living at present in widely separated and narrowly circumscribed regions ; *Ceratodus* (Fig. 21) is known only from the Burnett River in Queensland, *Protopterus* inhabits the Gambia in North Africa, and *Lepidosiren* the reedy swamps of the Gran Chaco in tropical South America. The swamps of the Gran Chaco, at the level of the southern tropic, support a dense vegetation of papyrus and reeds ; the stagnant water in the wet season varies from about four to eight feet in depth, and an occasional sluggish stream meanders round the bases of the papyri and swamp-grasses. In the dry season the standing water may almost completely disappear. *Lepidosiren* is a sluggish fish, creeping about among the vegetation with the help of its hind fins, which are employed almost as legs. At regular intervals it visits the surface of the water to expire the stale air from its lungs and to inhale a fresh supply ; the frequency with which these visits to the air are paid depends on the condition of the water, and in very stagnant and foul water the intervals between the breaths may be only four or five minutes. It is clear therefore that even in the wet season, when water is comparatively abundant, the lung acts as an important accessory to the gills in the function of respiration. Its most essential function, however, is exercised in the dry season. Before the onset of this season the *Lepidosiren* feeds actively on the molluscs and other animals in the swamp, but as the drought approaches feeding ceases, the animal burrows into the wet mud and fashions a kind of cocoon for itself with a small opening to the air protected by a lid. Curled up in this cocoon with its tail folded over its

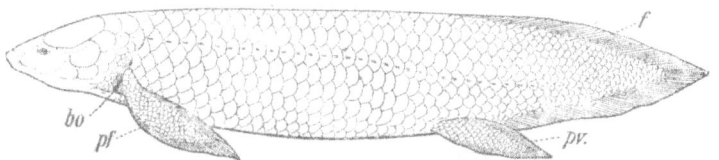

Fig. 21. *Ceratodus forsteri*. *f*. median fin; *pf*. pectoral fin; *pv*. pelvic fin; *bo*. branchial opening. (After E. S. Goodrich, Lankester's *Treatise*.)

face, the *Lepidosiren* secretes a mass of mucus round itself to prevent too great desiccation; it^ breathes entirely by means of its lungs, and awaits the return of the rains to float it out of its cocoon. As soon as the floods liberate it, *Lepidosiren* proceeds directly to the duties of reproduction, and another, though differently constructed, nest is built in the mud below the water to receive the eggs and rear the young.

The habits of the African Lung-fish, *Protopterus*, are very similar; it too burrows in the mud on the approach of the dry season; but the Australian Lung-fish, *Ceratodus*, appears to differ in this respect from its two relatives. The air-breathing operations of

Ceratodus are exercised in exactly the same way as in *Lepidosiren* ; the animal coming to the surface of the water to breathe at regular though longer intervals. *Ceratodus*, however, never burrows in the mud, and during the dry season it continues to live in the ordinary way, seeking out the deeper pools which are not liable to dry up completely. The advantage which *Ceratodus* obtains from its lungs is that it can subsist in foul water for a much longer period than an ordinary fish, which depends on its gills alone for respiration.

The earliest known land Vertebrates with typical pentadactyle limbs, the Stegocephali, make their appearance in the lower Carboniferous strata ; the Dipnoi are geologically much older, going back in much their present form to the Devonian and probably still older Palaeozoic formations. *Ceratodus*, in the identical shape in which it now exists in Australia, is known as a fossil from the Trias of Europe, so that this fish, like the Crustacean *Anaspides*, is one of the most interesting of persistent types, with a long geological history, known to us. The extreme antiquity of the Dipnoi favours the view that they are closely related to the true ancestors of the land Vertebrates, and this is further confirmed by resemblances in their anatomy and development which are not clearly associated, like the lungs, with their present habits or mode of life.

It is especially in the structure of the skull and its relation to the jaws that the Dipnoi differ essentially from all fishes and resemble the Amphibia or lowest order of land Vertebrates. In the ordinary fish the upper jaw is not fused to the skull, but is loosely attached to it, the most important attachment being effected by the upper portion of the first gill-arch, the hyomandibular cartilage or bone, which is firmly fixed to the back of the skull near the internal ear, and is at the same time attached to the upper jaw near the articulation of the latter with the lower jaw. This suspension of the jaws on the skull by means of the hyomandibular is termed hyostylic (Fig. 22 A), and it is characteristic of all fishes except the Dipnoi. In these fishes the upper jaw is completely and immovably fixed on to the skull, so that, as in the land Vertebrates, the Dipnoi cannot move their upper jaw without moving the whole skull. The hyomandibular cartilage, which no longer takes any part in the support of the jaws, is reduced to a minute functionless rudiment or has entirely disappeared. We thus see that the Dipnoi are related to the land Vertebrates, not only by the possession of similar lungs, but also in the fundamental character of their jaws and skull. The spiracular cleft, i.e. the gill-cleft in front of and corresponding to the hyoid arch, is, as we have already mentioned, functional only in the Selachians;

in the Bony-fishes and Dipnoi it is functionless and does not open to the exterior. In the passage from the water to the land the Vertebrate has undergone a remarkable and important change in respect to the spiracular gill-cleft and the hyoid arch behind it, these structures being brought into the service of an entirely new organ, the sense organ of hearing (Fig. 22 B). Animals living in the water have no use for an organ of hearing, and in fact such an organ is unknown in water animals of all kinds, the only animals which can hear being the land Vertebrates and certain terrestrial insects. The ear of the fish, consisting of certain canals buried in the wall of the skull, is not an ear at all in the sense of being an auditory organ, its function being purely that of balancing. The land Vertebrate, beside possessing the original balancing organ of the fish, or internal ear, has developed an accessory mechanism for collecting and transmitting sound-waves in the air to a specialized part of the internal ear, which is supplied with a branch of the eighth cranial nerve and is sensitive to the sound-waves. The accessory mechanism developed by the land Vertebrate consists in an external ear, which may be absent or rudimentary, a tympanum or ear-drum on which the sound-waves strike, and a middle ear or cavity through which the sound-waves are carried to the internal ear by means of a chain of small bones, the auditory ossicles, which are attached by one end to the ear-drum and by the other to the internal ear. Now this middle ear cavity is developed in the land Vertebrate from the spiracular gill-cleft which no

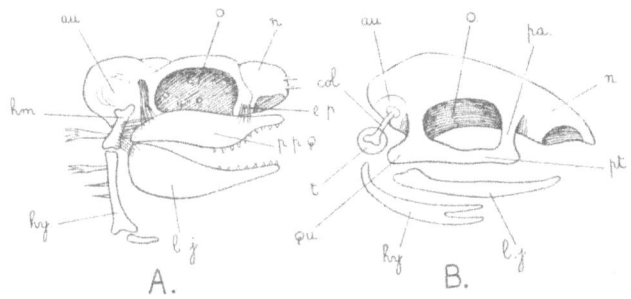

Fig. 22.   A.   Hyostylic skull of a Selachian, e.g. a Dogfish, in which the upper jaw or palato-pterygo-quadrate bar *p.p.q.* is attached loosely to the cranium by means of the ethmo-palatine ligament *e.p.* and by ligaments attaching it to the hyomandibular cartilage *hm.   au.* auditory capsule; *hy.* ceratohyal cartilage; *l.j.* lower jaw; *n.* nasal capsule; *o.* orbit.   B.   Autostylic skull of a terrestrial Vertebrate, e.g. a Frog, in which the upper jaw, composed of the quadrate, palatine and pterygoid, is firmly fused with the skull.   The hyomandibular cartilage has been transformed into the columella auris (*col.*) which conveys the soundwaves from the tympanum *t.* across the middle ear to the internal ear situated in the auditory capsule *au.*

longer breaks through to the exterior but is closed by a membrane which now forms the tympanum or drum of the ear.   Internally the middle ear opens into the throat by the Eustachian tube, which is the old opening into the throat of the spiracular gill-cleft. The hyomandibular cartilage, which in the fish was used to suspend the jaws to the skull, has been pressed into the service of the middle ear, and in the lower forms of land Vertebrates, the Amphibia, Reptiles and Birds, supplies the auditory ossicle for conveying the sound-waves from the drum of the ear to the internal ear (Fig. 22 B). In the Mammal further changes have occurred ; not only does the hyomandibular cartilage serve as an auditory ossicle, the stapes, but additional ossicles, the incus and malleus, are present which have been segmented off the jaw articulation and correspond to the quadrate and articular bones of the lower land Vertebrates.

The rest of the gill apparatus, including the body of the hyoid arch and all the succeeding arches, though formed as separate and distinct arches in the embryo land Vertebrate, are fused together in the adult to form the hyoid cartilage which supports the throat and vocal apparatus, so that all traces of their gill nature is lost, and were it not for their embryonic history which involves not only themselves but the blood-vessels, nerves and muscles, associated with them, we should be hardly able to prove the un-doubted origin of the land Vertebrate from a fish-like ancestor. That this has occurred is, however, certain ; the gradual embryonic and larval changes

of the Amphibian, such as a Frog, from the fish-like
Tadpole to the terrestrial air-breathing adult afford
a picture of how the change has taken place ; the
higher land Vertebrates, which never pass any of their
life in the water, have preserved in their development the abbreviated history of their gill-possessing
ancestors ; and the Dipnoi, both in the nature of
their lungs and in their skull, betray their relationship to the primitive group of fishes which gave origin
to the first land Vertebrates.

CHAPTER VII

THE RISE OF THE MAMMALIA

THE Mammals, comprising the most highly
organized animals with man at their head, are
characterized by a combination of features which
sharply distinguish them from their nearest living
relatives, the Reptiles. Among the more important
of these features are, the possession of a covering
of hair in place of the reptilian scales ; associated
with this character is the warm blood, kept at a
constant temperature ; the teeth arise in at most
two successions, whereas in Reptiles they may be
replaced an indefinite number of times, and whereas
in the Reptiles the teeth are all simple cones, in the
Mammal they are differentiated into several different
kinds, incisors in front for cutting, canines for de-
fence, premolars and molars for grinding. In the
skeleton a large number of peculiarities might be
enumerated. The vertebrae are divided into a constant number of cervical, dorsal, lumbar and sacral
vertebrae ; the skull articulates with the vertebral
column by two occipital condyles, instead of one ;
the brain cavity is greatly enlarged, a sign of superior

mental powers ; the articulation of the upper and lower jaws has shifted from the quadrate and articular bones to the squamosal and dentary. But in the Mammals, the characteristics which are of the greatest interest and will principally engage our attention, concern their method of reproduction and of nourishing their young. It is a significant fact that in the three great groups of animals in which mental and social development have reached their highest pitch, namely, Insects, Birds and Mammals, we find that the care and nurture of the young are carried to a high degree of perfection, and occupy a very large share of the parents' activities. This parental care for the offspring forms the basis, first for family life and then for the formation and sustenance of larger and more complicated societies, and the social habits so engendered encourage the development of the higher mental faculties, which we seldom if ever meet with among the lower and more individualistic creatures which take no further interest in their eggs after they have been laid and for whom parentage implies no ulterior duties. We are therefore justified in paying special attention to the peculiar modes of reproduction and nurture in the Mammalia, as without their attainment social man could hardly have been produced, and with their decay, permanence or change the ultimate fate of man is intimately bound up.

The origin of the skeletal characters of the Mammalia and of their teeth—in fact, of all those hard parts which are capable of being preserved in fossilized remains—is clearly indicated to us in a group of ancient and extinct Reptiles, the Anomodontia, which flourished in the Secondary period. Their remains have been found in Triassic strata in widely separated parts of the earth, viz. in S. Africa, in Scotland and in Eastern Europe, and they afford a beautifully clear transition from the reptilian to the mammalian type of structure. These Anomodont Reptiles had not only acquired the essential characters of the mammalian skeleton and teeth, but they had branched out into various forms, in accordance with various modes of life, which simulate remarkably the divergent groups of modern Mammals, e.g. some were carnivorous, others herbivorous, others insectivorous, and these several habits have left their indubitable mark, especially on the characters of the teeth in these early mammalian ancestors. But on internal structure and mode of reproduction these fossils throw little if any light.

It is a very familiar fact that Birds and Reptiles lay eggs, that these eggs are incubated either by the parents or by the action of the sun or the heat engendered by the fermentation of decaying vegetable matter, and that from the eggs the young emerge and may straightway care for themselves, or, especially in the case of Birds, be supplied with food collected by the care of the parents. The egg of the Reptile or Bird is relatively very large, and its size is due to the great amount of yolk which it contains. This yolk is the food material on which the embryo subsists during its development; it supplies the growing embryo with the material from which it

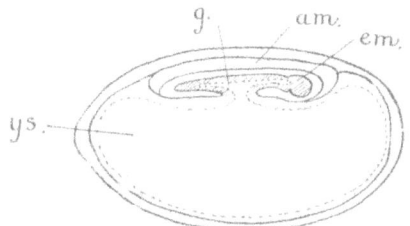

Fig. 23.   Longitudinal section through a Bird or Reptile embryo in its egg-shell.   *am.* amniotic cavity; *em.* body of embryo; *g.* gut of embryo communicating by the umbilicus with the large yolk-sac *ys.*

may build up all the various organs of its body. The embryo itself, as it develops, floats, so to speak, upon the yolk, from which it becomes folded off, but to which it remains attached at one point, the umbilicus (Fig. 23).   The umbilicus attaches the yolk-sac to the growing embryo, and it is in reality a hollow stalk by which the yolk-sac is in communication with the hind part of the alimentary canal of the embryo, so that the yolk can pass into the gut of the embryo and so be used as food. Numerous blood-vessels arising from the umbilical veins and arteries pass out of the embryo and spread over the yolk-sac, and by their assistance the food material of the yolk is taken up into the general circulation of the embryo and distributed all over its body. As the embryo increases in size and its organs become larger and more completely developed, the yolk-sac becomes included in its alimentary canal, and a few days after hatching it is completely digested, and the young reptile or chick must be supplied with food from without.

In this process development takes place within the egg-shell entirely at the expense of the yolk, which has been stored by the parent in the egg before fertilization.

In the case of a typical Mammal, such as a rabbit, a dog, or a human being, an entirely different mode of foetal nutrition takes place. The egg when it leaves the ovary is exceedingly small and contains a minimal quantity of yolk. On attaining to the oviduct and being fertilized, it passes into the uterus, where it becomes attached and finally embedded in the uterine wall. Here it begins to develop, and it is at first nourished by the secretion of glands in the uterine wall of the parent.

It is necessary, in order to follow how the

nutrition of the foetus is now brought about, to understand the principal features in the development of the foetus itself (Fig. 24 A and B). The repeated division of the fertilized ovum results in the formation of an embryo, only part of which is destined

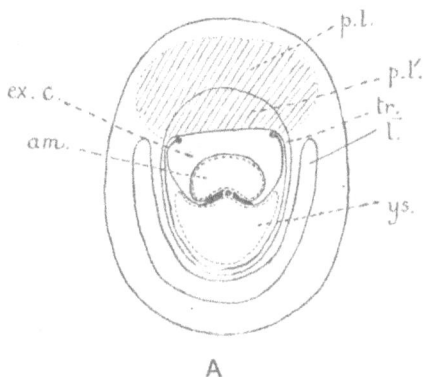

A

Fig. 24.   A.   Transverse section through the uterus of a Mammal in the wall of which is embedded an embryo at an early stage before the formation of the placenta. The region where the placenta will form is shaded in.   *am.* amniotic cavity; *l.* lumen of uterus; *tr.* trophoblast; *ys.* yolk-sac; *ex. c.* extra-embryonic coelom; *pl.* maternal portion of placenta; *pl'.* foetal portion of placenta.

to form the future animal.   The external investing layer of cells, known as the trophoblast (*tr.*), is simply used to put the foetus inside it in close connection with the maternal tissues and so to convey the

nourishment from the mother to the young. Within
the investing trophoblast the cells continue to divide

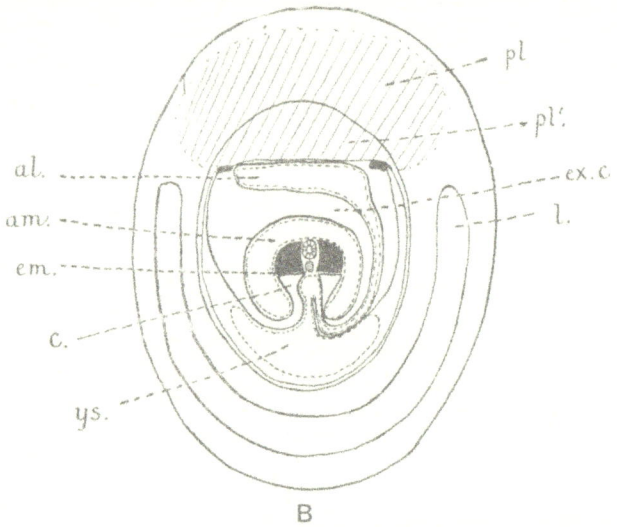

B

Fig. 24. B. Transverse section through the uterus of a Mammal and a developing foetus after the allantoic placenta has
formed. The foetus *em.* is pushed up into the amniotic cavity
am. and is attached to the yolk-sac *ys.* by the umbilicus, and to
the placenta *pl.* by the allantois *al.* The coelom of the foetus *c.*
is continuous with the extra-embryonic coelom *ex.c.* The placenta
is formed partly of uterine *pl.* and partly of trophoblastic tissue
pl'.

and to arrange themselves into the three germ-layers
of the embryo—the ectoderm, endoderm, and mesoderm. Just as in the Reptile or Bird, the endoderm

forms a rather large yolk-sac, which extends outside
the region from which the foetus is actually formed ;
so that as in a Reptile the embryo floats, so to speak,
on its yolk-sac. But this yolk-sac is much smaller
in the Mammal than in the Reptile or Bird, and
instead of containing yolk, it contains a small
quantity of albuminous material. The yolk-sac,
however, plays some part in the nutrition of the
embryo after the latter with its investing trophoblast
has been embedded in the uterine wall. The trophoblast becomes thickened at this stage ; at the same
time the uterine tissues of the mother become
highly charged with blood, in consequence of which
they swell and the activity of the uterine glands is
greatly increased. The tissue of the embryonic
trophoblast becomes closely adherent to or may
even completely fuse with the uterine tissues to
form a common mass in which the origin of cells
from the parent or embryo can no longer be deter-
mined. While this condition of things is being
established, the yolk-sac, which is applied to the
trophoblast over a large part of its area, absorbs
nourishment from the trophoblast and conveys it to
the growing foetus lying on its surface. As soon,
however, as the fusion of trophoblast with the
maternal uterus is completed, which happens with
considerable speed, the yolk-sac ceases to be an important organ of nutrition, its place being taken by

8—2

a new organ, the allantois (Fig. 24 B *al.*). At the time
that the allantois grows out from the foetus the latter
is being folded off from the yolk-sac. The embryo is
covered over with an investing envelope, the amnion,
and between this envelope and the yolk-sac there is
a spacious extension of the embryonic coelom to
form an extra-embryonic coelomic space. It is into
this space that the allantois grows out from the hind
gut of the foetus near the region of the umbilicus,
pushing with it the coelomic lining of the part of the
gut from which it comes.

Accompanying the allantois, in the coelomic
mesoderm, blood-vessels grow out; the allantois
fuses with the thickened trophoblast where the latter
has already amalgamated with the uterine tissue,
and its blood-vessels vascularize the common mass,
and come into the most intimate relation with the
maternal blood-vessels. This highly vascular mass
of tissue, composed partly of the swollen uterine
wall of the mother and partly of the embryonic
trophoblast, is vascularized on the one hand by the
maternal blood-vessels, on the other by the foetal
blood conveyed in the allantoic vessels. The com-
posite organ so formed is known as the placenta
(Fig. 24 B *pl.* and *pl'.*), and it is by this organ that
the real business of the nutrition of the foetus
during gestation is carried out. When the allantois is fully established, with its blood circulation

conveying the nutriment from the placenta to the foetus, the yolk-sac ceases to function and dwindles, the foetus lying in the amniotic envelope becomes completely folded off from the yolk-sac save at the umbilicus, and it is attached to the uterus of the mother by the allantoic stalk.

It was for a long time a matter of dispute as to whether the blood of the mother actually mingled with that of the foetus in the placenta. In some forms of Mammals where the fusion of embryonic trophoblast and maternal tissue is very intimate there can be no doubt that the maternal capillaries pour their blood into spaces in the placenta which actually communicate with the capillaries of the foetus. In other cases where there is no absolute fusion of the maternal and foetal tissues, but only a close adherence or interdigitation, it seems equally certain that all the interchange of fluid material is by diffusion from the maternal to the foetal circulation, not by commingling. This is, however, a point of not very great importance. Certain it is that the vascular connection between parent and foetus during gestation is a very close one, and that various bacterial diseases and physical states can be transmitted from parent to offspring during this period so as to simulate the appearance of the inheritance of acquired characteristics.

During gestation the placenta stores quantities

of carbohydrate and fatty material, which is conveyed to the foetus in the blood as occasion requires; proteid and oxygen are supplied by the maternal blood, while excretory matter is also removed by the maternal circulation.

There are great variations in the form of the placenta in the different groups of placental Mammals. Various classifications have been attempted, but the subject is a complicated one and by no means settled. In the Indeciduate Mammals, e.g. Ungulates and Lemurs, the maternal and embryonic tissues do not come into inseparable connection, so that at birth the foetal part of the placenta formed from the trophoblast comes away from the maternal portion, which is left behind and subsequently absorbed *in situ.* In the Deciduata, including the great majority of Mammals, e.g. Carnivora, Rodentia, Insectivora, Primates, the fusion of maternal and embryonic tissue in the placenta is complete, so that at birth the whole placenta is shed as the "after-birth," and no separation of maternal and embryonic placental tissue is possible. Besides the differences of the Deciduata and Indeciduata, there are also differences in the way in which the trophoblast of the embryo is attached to the uterine tissue, resulting in the formation of different kinds of placenta, e.g. diffuse or zonary. In the Primates the mode of placentation is peculiar, because the vascularization of the placenta, which is formed all round the trophoblast, is not brought about by the allantois but by a mesodermal stalk developed much earlier. We cannot, however, enter here into the various details of this subject, many of which are difficult to observe and still more obscure to interpret.

The period of gestation varies considerably in the different orders of Mammals according to the size and state of development at birth of the young and to the number of young in a litter.

After birth the young Mammal is nourished in a characteristic way by its mother, namely, by suckling. The Mammals are characterized by the possession of peculiar skin-glands, some of which secrete sweat for the regulation of the body temperature ; others secrete sebum, an oily material for keeping the hair moist and pliable. Still others, probably homologous and of an essentially similar nature to the foregoing, are collected into special regions of the thorax or inguina and secrete the nutrient fluid, milk. These mammary glands, as they are called, open on to special papillae, the teats, from which the young after birth can suck a copious supply of milk. The number of teats possessed by the female of any Mammal is proportional to the greatest number of young generally produced in a litter. Milk constitutes a complete diet for the suckling young, containing carbohydrate, fat, proteid and salts in requisite proportion.

As suckling proceeds the young Mammal begins to prepare itself for a different kind of food, and it begins to cut its first or milk-teeth. We have already stated that the mammalian teeth are differentiated into incisors in front, a pair of canines at the side, and behind these a certain number of premolars and molars. Of these teeth only the incisors, canines and premolars are represented by milk-teeth, which are replaced by permanent teeth afterwards; the molars or grinders are not developed as milk-teeth which are subsequently replaced, but they appear for the first time as permanent teeth synchronously with the appearance of the other permanent teeth in front.

The reason why milk-teeth are developed by the young Mammal is that, at the period when suckling ceases and the animal has to masticate solid food, the head and jaws have not attained by any means to their full size and there would not be room either for the full complement of teeth or for teeth of so large a size as are required by the more powerful adult animal. Hence the young Mammal is first of all furnished with a set of small milk-teeth without any molars at the back, which are gradually replaced as the head and jaws begin to attain their adult size, the molars being added at the back of the jaws as the last of the series.

We see, therefore, that in the typical Mammal three sets of organs are concerned in the peculiar mode of reproduction: the placenta and its associated structures, the mammary glands, and the teeth.

There exist, however, two groups of Mammals, the Monotremata and the Marsupials, which possess these characteristics either not at all or else in a very much simplified form, and though it may be true that this simplification is due to some extent, especially in the Marsupials, to a secondary loss or degeneration, yet there can be no doubt that these animals have left the main Mammalian stem at a period before the reproductive arrangements had reached their present degree of perfection, and that they have retained to some extent a primitive condition when these arrangements were in course of evolutionary development.

The Monotremes are represented by two very peculiar genera of animals, the *Echidna* and *Platypus* (see Frontispiece), which exist to-day in Australia, Tasmania and New Guinea, but nowhere else in the world ; nor do they occur as fossils save in the recent deposits of these countries, so that their geological history is really unknown, though we must suppose it to be a very ancient one. They possess the essential structural characters of Mammals ; thus they are clothed with hair and they suckle their young with milk ; the skeleton is essentially Mammalian, though with certain Reptilian reminiscences, especially in the limb girdles, and a Reptilian condition is still preserved in the fact that the apertures of the rectum and of the urino-genital system open into a common cloaca, whereas in the higher Mammals these apertures are always separate.   It is, however, in the mode of reproduction that the Monotremes have retained the most undoubted traces of their Reptilian ancestry, since, instead of nourishing the foetus in the uterus by means of a placenta, they lay eggs which they hatch out in a nest.

The *Echidna*, known to the Australian settlers as the Native Porcupine, is a small animal, sometimes attaining two feet in length ; it is covered with sharp quills and has the power of rolling itself up into a prickly ball when it is in fear of an attack.   The jaws are prolonged into a very long snout, and the tongue is long and extensile, these characters being adapted to secure the ants on which the creature subsists.   The limbs are very stoutly built and fashioned for digging, since the animal not only must burrow for its prey, but it also excavates a hole in the ground, usually under a rock or tree-stump, where it makes its nest. The female *Echidna* deposits a small egg with a leathery shell, resembling that of a Reptile ; it is transferred, after being laid, to a small pouch on the under surface of the female which develops only at

the breeding season.   Here, after a very short incu-
bation, the young *Echidna* hatches out in a very
undeveloped and helpless state, and it proceeds to
suck the milk of the mother which is secreted by
mammary glands, the openings of which are scattered
over the surface of the breast in the region of the
pouch.   In this situation the young *Echidna* continues
to develop, until it is ready to go out and seek its
insect food for itself.   At no time of its life does the
Echidna develop any teeth, though the palate and
tongue are covered with horny spines for masticating
the bodies of the ants on which it feeds.

The *Platypus*, which is really very closely related
to the *Echidna*, follows a very different mode of life,
being amphibious in its habits and frequenting rivers
and mountain lakes where it feeds upon fresh-water
molluscs and the fry of fish.   Instead of being covered
with quills, it is clothed with a very rich and smooth
fur, something like an otter's ; its toes are webbed
and it possesses a flat horny bill very much like a
duck's, with which it catches the shell-fish.   The jaws
are furnished with horny plates in place of teeth for
grinding the hard shells, but this animal during
development exhibits the rudiments of true tuber-
culated teeth in its jaws which are never cut to the
surface, thus proving that these animals possessed in
the past the Mammalian type of teeth.   The Duck-
billed *Platypus* excavates a burrow in the bank of

a lake or stream, one opening of the burrow being above the level of the water, the other below, so that it can enter its home either from the water or the land.   In the breeding season the female lays a small yellowish egg, which it incubates in a rough nest of grass or sticks at the back of the burrow ; here the young hatches out and is suckled by the parent.   No pouch develops in the *Platypus*.

The Monotremes illustrate very well the limitations with which the term Primitive Animal must be accepted.   In the fact of their laying eggs and in some of their anatomical features these animals undoubtedly retain certain Reptilian features which we know on other grounds the ancestors of the Mammalia must have possessed, but we should be very far astray if we figured the Mammalian ancestors in anything like the shape of the two existing Monotreme animals.   These two animals in many respects are highly specialized forms which owe many of their characteristic features, such as the loss of teeth, the curious structure of the mouth and jaws, the webbed feet and the limbs adapted for digging, to peculiarities in their habits, which are probably modern acquisitions and were certainly not possessed by the generalized ancestral form which gave rise both to them and the modern placental Mammals.

Whilst the Monotremes may be certainly classed as Primitive Animals, with the restrictions mentioned, there would appear to be rather more doubt whether the Marsupials have a right to the same title or whether their comparative simplicity in the matter of reproduction is due to degeneration. There can be little doubt that the Marsupials have degenerated to some extent from a more advanced Mammalian stock, but the balance of evidence would seem to indicate that at the period when degeneration set in they had by no means attained to the high standard of development possessed by the modern placental Mammals.

The great majority of the Marsupials at the present time inhabit the Australian region, with the exception of two forms, the American Opossum and the American Bush-rat ; fossil Marsupials are, however, fairly abundant in America and even in Europe, and among these fossil forms some are of very great interest owing to their great antiquity. The fossil jaws from the Stonesfield Slate (Jurassic Period) are known to have belonged to animals closely allied to modern Marsupials, and they are the earliest known remains of an undoubted Mammalian nature, since the formation in which they occur is of Secondary age, whereas all other undoubted Mammalian fossils are Tertiary. These fossil jaws belonged to animals of an insignificant size, but in Australia the Marsupial stock has diverged into a great variety of stems and includes forms as small as mice, as large as the rhinoceros, and as surprising as the kangaroo. The reason of this great development of the Marsupials in Australia is not far to seek. Australia, at any rate since the beginning of Tertiary time and probably long prior to that period, has been entirely cut off from the rest of the world, so that no representatives of the higher Mammals, save man, his dogs, and a few species of rats, have been able to colonize it. The Australian and Tasmanian natives themselves probably did not reach Australia until very late in Tertiary time, and they have never developed sufficiently to act as an exterminating force on the wild animals. Hence on the vast Australian Continent and the connected islands of New Guinea and Tasmania, the Marsupial fauna has enjoyed an asylum free from competition with the more perfect types of Mammalian creation, which were being evolved in the rest of the world and were rapidly supplanting the Marsupials there with which they came in contact. The manner in which Australia first received its Marsupial fauna is a matter of speculation, but there is much to be said for the view that they reached that country at a time when it was connected with South America by arms stretching up from the Antarctic Continent, at a period previous to the development of any of the modern Mammalian types. Secure in the isolated Australian region, the Marsupials are now represented by types adapted to live under every sort of condition ; the

Kangaroo and the gigantic extinct Wombat, *Diproto-
don,* follow, or used to follow, a herbivorous habit; the
Phalangers, misnamed Opossums owing to a spurious
resemblance to the American Opossums, live in trees
and feed on their foliage, while some have developed
parachutes and glide from branch to branch like
flying-foxes or squirrels; the Dasyures or native cats
are small carnivores, which prey chiefly on birds, and
there are larger carnivorous Marsupials which now
only survive in Tasmania, the Tasmanian Tiger and
the Devil.   This does not by any means exhaust the
variety exhibited by the Australian Marsupials; some
are of a minute size and live like mice; others burrow
in the ground and have assumed a remarkable resemblance to moles; the Wombat feeds on roots and lives
like a badger; the Bandicoot (*Perameles*) is insectivorous or omnivorous.

But though these animals assume such various
guises and follow such a variety of habits, simulating
now one class of the higher Mammals and now another,
yet they are all characterized by certain peculiar anatomical and physiological features which are never
found in the higher or Eutherian Mammalia.   The
most important of these features concern the development and nutrition of the foetus, the suckling
of the young in a pouch, and the almost entire absence
of a milk dentition.

The young of the Marsupials are born in a very

undeveloped state while they are still exceedingly small; thus the newly born young of a kangaroo, which when adult measures upward of six feet high, is less than half-an-inch in length when first delivered by the mother, and it is still in the developmental stage of a very early foetus of one of the higher Mammals.   After delivery, the mother picks up the tiny foetus with her lips and transfers it to her pouch, placing the lips of the foetus on to one of the nipples which are enclosed in the pouch.   The lips of the foetus become firmly fixed to the nipple, and here it is suckled by the mother for a very long period until it is about a quarter full grown.   The discovery of the minute foetus attached to the nipple of the mother led to the idea, which is still current among the settlers in Australia, that the young are actually born in this situation, a supposition so grotesquely at variance with all the facts of Mammalian anatomy and physiology as hardly to deserve the minute and circumstantial refutation it has naturally received.

The birth of the Marsupial young in so small, helpless and undeveloped a condition, as compared with the young of an ordinary Mammal, is due to the comparatively inferior nutrition of the embryo while still in the uterus of the mother.   The Marsupial embryo is in fact half-way between the condition of the Monotreme, where the embryo is delivered as an egg and is entirely dependent on the yolk contained in the yolk-sac, and that of the Eutherian Mammal, which is born in an advanced state of development after being nourished in the mother by means of the allantoic placenta.

The Marsupial embryo is nourished in the uterus of the mother almost entirely by the yolk-sac, which contains food material itself and also comes into contact with the uterine wall of the mother and forms a poorly developed yolk-sac placenta. This mode of nutrition or gestation lasts for only eight days or so, when the foetus is delivered.

Now the remarkable fact is that in the Marsupial foetus an allantois is actually formed, growing out from the hind gut in a typical manner; but soon after its formation, instead of growing out and fusing with the trophoblast and the uterine wall, it begins to shrivel away and its blood-vessels atrophy and disappear. In only one Marsupial, the Bandicoot (*Perameles*), does the allantois succeed in reaching the wall of the maternal uterus and forming a small true placenta, which acts as a feeble source of nutriment for a few days (Fig. 25). It is evident that we are dealing here with a degenerate condition. The allantois begins to develop, but in the majority of cases it degenerates before it has performed any function at all. This certainly would not be the case if it were in a primitive condition, representing the origin of the placenta, because in that case it would at any rate succeed in getting into communication with the maternal tissues, whereas this only happens in a single case.

The evidence is conclusive that the Marsupials are descended from a form of Mammal which had a functional allantoic placenta, but whether this placenta was ever very much more highly developed than it is in *Perameles* remains open to question.

At birth we have stated that the small and little-developed foetus is transferred by the mother to her marsupium and affixed to one of the nipples there. The question arises as to whether the possession of a marsupial pouch is a primitive characteristic of the early Mammalia or whether it is a special acquisition of the Marsupialia. We may answer with confidence that this is a primitive character, firstly because a marsupium is possessed by the Monotreme *Echidna*, and secondly in the young of a great many of the Eutherian Mammals two ridges are developed on either side of the mammary glands representing the rudiment of a marsupial pouch which subsequently disappears.

The young of the Marsupials spends a very long time in the pouch, and it is characteristic of this class of Mammals that the suckling period is much more prolonged than in the higher Eutheria. In connection with this lengthened suckling period a peculiarity in the development of the milk-teeth is to be noticed.

Whereas in the Eutherian Mammals all the permanent
incisors, canines, and premolars are preceded by representatives in the milk dentition, in the Marsupials
only two pairs of teeth, viz. the last premolars in

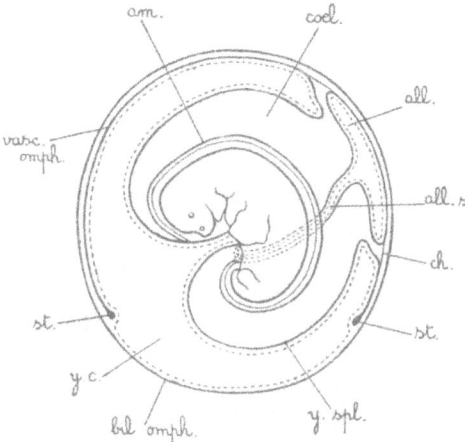

Fig. 25.   The foetus and embryonic membranes of *Perameles.   am.*
amnion; *coel.* extra-embryonic coelom; *all.* allantoic placenta;
all.s. allantoic stalk; *y.c.* yolk-sac; *vasc. omph.* vascularized
coelomic surface applied to yolk-sac; *st.* the sinus terminalis in
which blood-vessels ramify; *bil. omph.* the trophoblast. (After
Hill, *Q. J. M. S.* vol. XL.)

each jaw, have milk predecessors.   What actually
happens is this—during the suckling period the incisors, canines, premolars and molars are gradually
cut to the surface; then before the pouch is per-
manently left the last premolars are replaced by

corresponding teeth of a different form which become the permanent last premolars. It was for long disputed to which series the deciduous premolars and the permanent teeth belonged, whether the permanent incisors, canines and front premolars belonged to the milk or to the true permanent series, and whether the deciduous last premolar represented a milk dentition in the process of development or of degeneration. It has now been found that a complete series of teeth, corresponding to milk incisors, canines and premolars, are formed as rudiments, but only one pair of these in each jaw, the deciduous last premolars, are actually cut to the surface. The deciduous last premolar represents therefore a milk dentition, the rest of the members of which remain below the gums. The permanent incisors, canines, premolars and molars correspond therefore to the permanent series in other Mammals, while the milk dentition is degenerate, with the exception of the last premolars. We arrive, therefore, at the conclusion that the Marsupials are not only descended from a form of Mammal with a better-developed placenta, but also from a form which possessed a complete milk dentition, so that in these two respects the modern Marsupials are degenerate and not primitive. The fostering of the young in the pouch is, however, a primitive character which has been lost in the higher Mammals. The degeneration of the milk dentition in the Marsupials

would seem to be due to the very long period spent
by the young as sucklings in the pouch, making it
unnecessary for them to develop teeth at an early
period before the jaws have attained to anything like
their definite size. The loss of the placenta would
also seem to be correlated with the early period at
which the young begin to be suckled in the pouch.
These two peculiarities of the Marsupials, therefore,
the loss of the placenta and the loss of the milk dentition, appear to be connected with the prolongation
of the suckling period at both ends, the young being
transferred to the pouch when very little developed
and remaining as sucklings there until well-grown.
One cause of this lengthening out of the suckling
period may perhaps be sought in the absence of any
very serious enemies which might be liable to pursue
and attack the females when cumbered with young.
It is well known that in kangaroo-hunts if a female
carrying young of any size in her pouch is hard
pressed, she will fling the young away, showing that
they are a serious impediment to her flight. Whatever
may be the explanation, the fact seems clear that the
Marsupials have specialized along the line of increasing
the importance of the suckling period in the pouch,
and have given up the advantages of a placental
nutrition, and an early weaning of the young provided with milk dentition, whilst the Eutherian Mammals have increased the efficiency of the placental

nutrition, have given up the habit of carrying the young in a pouch, as being too cumbrous, and have accelerated the weaning of the young, which are early provided with a milk dentition.

The Marsupials have never attained to a very high grade of development, and their achievements have been practically confined to the Australian Continent and its adjuncts, Tasmania and New Guinea ; the Eutherian Mammals, on the other hand, with all the rest of the world as their field of activity, have branched out luxuriously into all the higher types of animals which we know ; the fleetness of the horse, the ferocity and strength of the lion and tiger, the strength of the elephant and the sagacity of man are the most finished and highest products of animal evolution.

All this latter development has taken place within the period of Tertiary time, and we can trace in the geological record the origin of all the great groups of the higher Mammalia from their common origin, the Condylarthra, in Eocene times. The unravelling of this history has been the triumph of palaeontological science and the vindication of the theory of evolution, but we are not concerned with it here, as the fossil forms which serve to link together the divergent groups have left few direct survivals, so that among the existing Mammalia we do not meet with any very striking examples of primitive animals.

It may have struck the reader that in our short and necessarily incomplete survey of the animal kingdom, we have been able to speak with far more confidence and certainty as to the course evolution has taken, when dealing with the Vertebrata than with the Invertebrata. This is to be expected on the theory of evolution, as the Vertebrata are the more modern products and there has been less time for extermination to wipe out the records of the connecting links. Within the limits of the Invertebrate phyla we are able to convince ourselves that a similarity of plan runs through the various components, but the incompleteness of the geological record makes the task of deciding as to which component is more or less primitive exceedingly difficult. We have seen that the phyla themselves are of immense antiquity ; that their origins go back probably far beyond the limits of geological time as reckoned by stratified, fossil-bearing rocks, so that the attempt to determine the relationships of these phyla must remain a matter of vague hypothesis or cautious scepticism. This limitation, however, need not shake our faith in the truth of the doctrine of evolution or of descent with modification, as it is inherent in the nature of the evolutionary process, being the outcome of the antiquity of life, of the prodigality of form in which life has been manifested, and of the vast scale on which extinction has taken place.

CHAPTER VIII

REFLECTIONS UPON THE PAST AND FUTURE OF ANIMAL LIFE

THE progress of biology, the acceptance of the doctrine of evolution and its application to man's antecedents have had a powerful influence in widening the horizon of his prospect upon the world, and we may applaud or deplore the partial detachment of the human mind from the ideal of perfection in another life to the hope of a progressive amelioration in this. The reasoning of a Darwin might inspire or give substance to the prophetic fancies of a Nietzsche, and the imposing spectacle of the march of evolution might open up an almost endless vista of progress before the hesitating feet of humanity.

But the enthusiastic vision, inspired by the certainty that evolution has taken place, must be subjected to the critical examination of how the evolutionary changes are accomplished, and in at least three particulars we must note a greater precision and distinctness of thought than was attainable in the middle of the last century, when the idea of evolution first seriously engaged the world's attention.

1. The distinction between those characters of an organism which it acquires by use or disuse during its life, or which are impressed upon it by its environment, and those characters which it receives as a birthright from its parents or which have originated in the germ from which it has sprung, was not clearly perceived until Weismann's teaching had taken root, but his central position is now the basis of all modern work on heredity, and has introduced a different temper into the believers of progress. Whilst it was still possible to hold that characters or attainments acquired during the life-time and activity of the organism were commonly transmitted to its descendants, a rapid and constant evolution in an upward direction seemed possible for the human race. Man had only to strive; his descendants would proportionally increase in virtue; and a race of men would be evolved which might know or even practise the proscriptions of the Mosaic dispensation from their earliest infancy. The realities of history and of heredity do not sanction such dreams, and we must be content to know that while man may lose almost everything by the loss of a tradition, he can never by vicarious effort spare his descendants the pain of assiduously acquiring it by practice. When we speak of the progress of man, what we are really interested in is his progress in the arts of civilization and refinement, but progress in these has

depended far more upon a continuity in tradition than upon anything analogous to the process of organic evolution.   Let these traditions once disappear, and the arts of civilization must be built up again from the beginning.

The recognition that permanent or hereditary advances in evolution arise not through an increased effort at attainment, but by an increased capacity for attainment which occurs quite mysteriously and apart from any exertion on the part of the organism, might direct our attention to the possibility of mankind advancing by this means, namely by the appearance of a race with a greater capacity for civilization in its widest sense, which would supplant existing races. This possibility is based on reason ; and there is no doubt that the process, which is the same as that which we believe to have operated in the evolution of organic life, has already played a considerable part in the history of civilized man.   As the conditions change under which men live, the type of man most fitted to survive and to leave offspring alters, so that man by consciously creating his own environment, which we call civilization, sets a standard by which certain types are selected for survival and others are eradicated.

Civilized races are on the whole less violent and deceitful, more intellectual and immune to zymotic diseases and to alcohol than savages, and they owe these qualities not solely and directly to the possession of good laws, education and hygiene, but partly because they have lived for a long series of generations in an environment of their own creation which has favoured certain types of individual more than others.   In short it is man's business to create his environment, but unless heaven sends his offspring more capacity, his labour is lost as far as evolution is concerned.   The growing conviction that organic evolution has taken place as the result of a selection of individuals exhibiting variations that have arisen mysteriously or, as we may say, by chance, has given rise to the school of Eugenists, who propose consciously to enforce the process of selection, if not for the survival of the fittest, at least for the eradication of the least fit.   It is impossible to estimate the influence which this school of thought may exert in the future ; the opinion of mankind may waver or decide as to the relative advantages of contracting marriages for reasons of sentiment or of public utility ; and the enactment of laws precluding lunatics and diseased persons from the privilege of propagating their kind might do more to stop the deterioration than actively to further the progress of the human race.

2.   The earlier ideas of evolution have been modified in another direction, and the dazzling picture of universal progress has been tinged with

the darker shades of decline and degeneration.   The
genius of Darwin may have realized the part which
degeneration has played in evolution ; but the proofs
of its widespread influence have accumulated on the
shelves of the learned without penetrating the minds
of the populace, for whom the words evolution and
progress are often identical.   The serious student of
evolution, relying upon the facts of comparative
anatomy and palaeontology, is forced to admit that
while the origin of the great advances in animal
organization are for the most part hidden from view,
the processes of extinction, degeneration and simplification are written large on all the great groups of
the animal kingdom as far back into geological time
as the record of them can be traced.   With all this
evidence before us, how can we be confident that the
present race of man is only a step in a continuously
upward process of evolution ?   If it is true that
without a stringent measure of selection, every part
and function of the organism is apt to degenerate or
at any rate is incapable of progressive improvement,
the environment in which civilized man exists would
seem to involve of necessity a decline or stagnation of
many of his bodily operations, while the progress of
his mental and moral powers is fostered solely by
tradition, almost unaided by that process of selection
which we must believe to be an essential condition
of any permanent advance in organization.   It is

often said that civilized man is no longer subject
to the processes of nature, and is extra-natural, but
the same condition has probably been reached over
and over again by organisms in a state of nature,
which have attained a certain equilibrium with their
environment when selection is practically in abeyance.
The condition is emphasized in man on account of his
immense powers of personal acquisition and adapta-
tion, which permit him within wide limits to imitate
the process of organic change by a counterfeit similar
in every respect save in the essential quality of
permanence.

3.   If we are in doubt as to the probability of
progress or decline in the future history of mankind,
our judgment may be still further perplexed by the
existence of a third alternative, which might receive
an equal sanction from history and reason.   It has
been our especial task in the course of these pages
to draw attention to examples of animal life still in
existence which have remained essentially unchanged
for vast though unknown epochs.   With the increasing knowledge of these persistent types has come
from many sides the conviction that the period of
time during which life has existed on the earth is
immeasurably greater than the earlier evolutionists
dared suppose, and that the process of organic change
may be incalculably slow and in many instances ar-
rested for an indefinitely long period.   The attempts

of *exact* science to correct the *inexact* speculations of biology, as to the age of the earth and the duration of life upon it, have ended in conceding to the biologist almost unlimited time for the accomplishment of the changes which he must believe to have occurred. We cannot readily explain the persistence of certain organic types unchanged through such untold ages; the reason may be found in the relation of the organism to its environment or in the untractable nature of the organism itself, which may in some way have lost the power of further variation and the acquisition of new capacities, but if the final congelation of the globe should discover the human race with the same average physical structure, and the same mental and moral capacity, as at the present time, the apostle of progress might sigh, but the philosopher would not hesitate, to record another example of a persistent type.

The progress, decay or stagnation of humanity may divert the speculations of a philosopher or engage the concern of a statesman ; man will continue to spread over the earth ; and a zoologist must enquire into the probable fate of those inferior creatures whom it is his especial province to study.

The primaeval savage, before he had acquired the rudimentary arts of civilization, maintained a dubious existence among the wild animals and in scenes of untamed vegetation, over which he exercised but little influence and could hardly aspire to rule. With the acquisition of his first rude weapons he waged a petty warfare against the large carnivores which threatened his safety, or against the herbivores which might supply him with food and raiment, and the bones of bears and tigers, of reindeer, oxen and mammoths are mingled with his own in the cave deposits and river drifts of the interglacial periods in Europe. At what stage in his development man first thought of domesticating such animals as the horse and dog, of herding cattle and sheep, and finally of bringing various fruits and vegetables under cultivation, we cannot exactly guess; but these first acquisitions of civilization go back to a period far antecedent to history, and few savages are so devoid of art as not to possess a single domestic animal. The Tasmanians, both when they were discovered at the end of the eighteenth century and at their extinction as a wild race a few years later, were devoid of this basis of civilization, and if they knew the use of language, of fire, and of stone implements, they could not boast a single domestic animal or cultivated plant. Their sustenance was gained by hunting unaided the wild kangaroos and opossums of the bush, or by diving for shell-fish, while their need for vegetable food was satisfied by a plentiful and natural supply of sea-weed and young fern shoots.

The assistance of the horse and dog in the chase, and the regular supply of food and raiment from cultivated plants and herded flocks, encouraged the rapid propagation of man over the habitable earth, and began that conquest of the world and subjection of its living inhabitants which continue with ever increasing velocity and widening scope at the present time. Within the period of history many interesting and beautiful animals have entirely disappeared from the face of the earth, and many more are threatened with extinction at no distant date. Most signal and conspicuous instances of extinction at the hand of man are afforded by the animal inhabitants of certain islands, which, until the advent of man, had dwelt undisturbed in an asylum from which, on the arrival of danger, escape was impossible. Of such animals, the Dodo has gained a pre-eminence and has passed into common language as the very symbol of extinction. The Dodo and the Solitaire were gigantic pigeons which existed solely in the islands of Mauritius and Rodriguez; their wings were atrophied and they had completely lost the power of flight, owing to the abundance of food and entire absence of any enemies. The European sailors who visited the islands in the seventeenth and eighteenth centuries effected the destruction of the helpless creatures, partly by slaughtering them for food, but chiefly by introducing pigs which ran wild and nourished themselves on the eggs of the birds.   The destruction of the Dodo and Solitaire by the introduced pigs is an instance of the common fact that man causes more havoc by unwittingly disturbing the balance of nature than by his immediate destructiveness.   The Island of St Helena, when first discovered in 1501, supported a dense vegetation of forest trees, among which the valuable Red Wood and Ebony were pre-eminent ; the forest protected a rich undergrowth and the soil was abundant and fertile.   At the present time the forest trees have disappeared and with their disappearance the soil has been washed away by the tropical rains, and the island is barren and desolate. This work of destruction was chiefly effected by the introduction of goats by the Portuguese in 1513, which ran wild over the island and browsed down the young seedlings ; the extermination of the forest was encouraged by the ruthless cutting down of the trees by the colonists, and the rich soil which was the work of geological epochs to create can never be replaced.

The Islands of New Zealand were inhabited by several different kinds of Moa, gigantic flightless birds related to, though much larger than, the Ostrich ; judging from the abundance of their remains they must have existed in large numbers, but their extinction was brought about by the Maoris who colonized New Zealand from the Polynesian Islands

or from Japan, and who still retained the tradition
of Moa hunts when they were discovered by the
Europeans in the seventeenth century. The advent
of the Maoris sealed the fate of the Moa, and the
Maori himself was to disappear before the invasion
of a race superior to himself in the arts of civilization
and destruction.

But the ever expanding activity of man has now
penetrated into the heart of every continent, and the
vast stretches of once unoccupied territory no longer
afford a safer asylum to wild animals than the oceanic
islands. The Savannahs of Australia have had their
population of kangaroo and opossums terribly thinned
since the occupation of the British, and the continent
of Africa south of the Sahara, which has harboured
in a safe retreat all the finest products of Mammalian
development, is losing its animal population with
astonishing rapidity. A picture of what South Africa
must have been a century or more ago is afforded by
parts of Uganda at the present time, where vast herds
of antelope, giraffe and zebra wander over the plains,
where rhinoceros and lions and great herds of elephant
are still abundant. But the British and Dutch in
South Africa have exterminated or at least decimated
the big game, and in East Africa inland from Mombasa
it has already been found necessary to form an artificial sanctum for the wild animals, and everywhere to
restrict the depredations of hunters. The preservation

of some animals would seem a superhuman task ;
the rhinoceros is too stupid to avoid and too ferocious
to fear a fatal encounter with man, and, while he is
unable to see, his sense of smell incites him to charge
indiscriminately upon single persons or upon caravans
of armed men who would often wish to leave him
unmolested. The elephant combines the qualities of
sagacity, destructiveness and usefulness, with an un-
tractability that may end in his final extinction. For
unknown ages he has served the peoples of India and
Ceylon as a beast of carriage, of transport and of
unrivalled strength ; no people in modern times has
had the skill or patience to tame the distinct African
species for domestic uses, and a believer in progress
might hesitate to ascribe to the Indian or the African
Continent the war-elephants which, in the times of
Hannibal, supplied the African and astonished the
Roman legionaries.   There is, however, no doubt
that Northern Africa in the second century B.C. could
boast not only elephants, but a people sufficiently
energetic to undertake the education of the colossal
beast.   But the elephant has ever shown itself
untractable to the full discipline of domestication ;
he may do everything in bondage at his master's
bidding save propagate his kind ; and the elephants
in India and in Ceylon, which may be seen carrying
their lords in the chase or engaged in the transport
of timber and other heavy burdens, have all been

10—2

born, nurtured and finally captured in the few fast-nesses of the wild jungle which still remain.

Looking forward into the future, we may see the tide of human expansion ever spreading, with here and there an occasional ebb, and in time all the living things that are not serviceable to man or cannot adapt themselves to the conditions he imposes will dwindle and disappear. The arts of breeding and acclimatization will produce new and curious artificial forms to supply the various needs of food, of raiment, of beauty and convenience which only life has the requisite cunning to produce and only human life the restless activity to require.

And with the universal disturbance in the economy of Nature caused by the migrations and activity of man, old resources of life and profit will be exhausted and new resources sought ; old diseases will die out or revive in places where they have long slumbered ; new diseases will develop as the teeming populations of man with their domestic animals and plants offer new opportunities to parasites to attack them.

With these problems and amid these surroundings the naturalist of the future will be engaged, and he may regret the time when the sea and land still contained undiscovered forms of life that might reveal little suspected links in the history of crea-tion, unexploited regions where the traveller might feel the thrill and the mystery of untouched and

unreflecting Nature. These times are passing, if they
have not yet wholly passed, away.   But the lover of
Nature may find comfort in the thought that man's
power over Nature is limited, and though robbed of
much of her wild beauty, she rises partly tamed but
unconquered in the end by the less-enduring spirit of
man.   In Italy, where the influence of civilized man
has been most intensely felt for long and continuous
periods, almost every feature in the landscape owes
its distinctive characters to man, and yet Nature has
assimilated and welded them into a harmonious
whole.   The cypresses, the pines, the dull-grey olives
on the brown hillsides, the vines, the oranges gleaming
like balls of yellow fire in the black foliage, all these
things that help to make Italy what she is were
brought hither and planted by man in distant ages ;
the native vegetation only holds its own, a stunted
remnant, on the mountain tops ; but who can regret
the change or say where the work of man begins and
that of Nature ends ?

APPENDIX A

CLASSIFICATION OF THE CHIEF ANIMAL PHYLA

Grade PROTOZOA. The body consists of a single cell, or of a few cells not differentiated to form tissues.

Grade METAZOA. The body consists of numerous cells aggregated together to form tissues.

Phylum COELENTERATA. The body consists of two primary layers, the ectoderm and endoderm (diploblastic). Instances —Jelly-fish, Anemones, Corals, etc.

Phylum PLATYHELMIA. The body consists of three primary layers, ectoderm, endoderm and mesoderm (triploblastic); no vascular system or perivisceral coelom. Body flattened dorso-ventrally. Instances—Turbellarians, Liver-flukes, Tape-worms, etc.

Phylum NEMATODA. Triploblastic animals, without vascular system or perivisceral coelom. Body round. Instances— Ascaris, Rhabditis, etc.

Phylum NEMERTEA. Triploblastic animals with vascular system but no perivisceral coelom. Body round. Instances— Cerebratulus, etc.

Phylum APPENDICULATA. Triploblastic animals with vascular system and typically a perivisceral coelom which may be secondarily reduced. Body usually metamerically segmented. Nervous system ventral.

Series *Annelida*. Marine, fresh-water or land worms with segmented bodies.

Series *Arthropoda.*
   Class 1. *Peripatoidea.* Peripatus.
   — 2. *Myriapoda.* Centipedes, etc.
   — 3. *Insecta.* Insects.
   — 4. *Trilobita.* Trilobites.
   — 5. *Arachnida.* Spiders, Scorpions, Mites, etc.
   — 6. *Crustacea.* Crabs, Prawns, etc.

Phylum MOLLUSCA. Triploblastic animals with vascular system and reduced coelom. Body not metamerically segmented, usually enclosed in a shell. Shell-fish.

Phylum ECHINODERMATA. Triploblastic animals with rudimentary vascular system and complicated coelom. Radial symmetry has been imposed on an original bilateral symmetry. Star-fish, Sea-urchins, etc.

Phylum BRACHIOPODA. Triploblastic, coelomate, unsegmented animals, enclosed in a bivalve shell. Lingula, etc.

Phylum ENTEROPNEUSTA. Triploblastic, coelomate, partially segmented animals, with proboscis and gill-slits. Instance— Balanoglossus.

Phylum CHORDATA. Triploblastic, coelomate, metamerically segmented animals, with dorsal tubular nervous system, and gill-slits.
   Division *Cephalochorda.* Amphioxus.
   Division *Urochorda.* Tunicates or Sea-squirts.
   Division *Vertebrata.*
     Class 1. *Pisces.*
     — 2. *Amphibia.*
     — 3. *Reptilia.*
     — 4. *Aves.*
     — 5. *Mammalia.*
       Subclass 1. Monotremata or Prototheria.
           — 2. Marsupialia or Metatheria.
           — 3. Placentalia or Eutheria.

APPENDIX B

A TABLE OF THE CHIEF GEOLOGICAL EPOCHS

(The figures indicating the relative depth of the strata are only roughly approximate. They are slightly adapted from Sir Ray Lankester's book on Extinct Animals.)

Period	Typical Formation	Characteristic fossils
Archaean or *Pre-Cambrian* 50,000 ft.	Schists and metamorphosed rocks, probably representing stratified rocks equal in thickness to all subsequent periods put together	No trace of fossils, the remains of which have all been destroyed by alteration of rocks
Cambrian 25,000 ft.	Welsh slates, etc.	Lingula, Trilobites, Nebaliid Crustacea, various Molluscs
Silurian 7,000 ft.	Slates and shales in Wales, etc.	Arachnids, Ostracoderm fishes
Devonian 5,000 ft.	Red Sandstones of Devonshire, etc.	Eurypterids, Elasmobranch and Bony fishes—(Actinopterygii and Dipnoi)
Carboniferous and *Permian* 15,000 ft.	Coal measures of England, sandstones and limestones	Trilobites dying out. Decapod Crustacea, Anaspids, Insects, Stegocephalian Amphibia. First appearance of terrestrial vertebrates

Palaeozoic (bracketed group: Cambrian, Silurian, Devonian, Carboniferous and Permian)

Secondary or Mesozoic	*Triassic* 3,000 ft.	White Lias of Central English plains, Elgin sand-stones, Karoo for-mation, S. Africa	Crinoids, Ammonites abund-ant. Anomodont Reptiles, and early Dinosaurs, Ichthyosaurs, etc. Earliest Mammalia
	Jurassic 5,000 ft.	Purbeck beds, Ox-ford Clay, etc.	Numerous early Mammalian remains. Reptiles highly de-veloped. Birds appear, Archae-opteryx
	Cretaceous 3,000 ft.	Chalk downs, etc.	True Birds, Bony fishes (Tele-osts). Many modern forms of Echinoderms and Molluscs.
Tertiary or Cainozoic	*Eocene* 1,000 ft.	London Clay, Paris beds, Puerco and Wasatch, Ameri-can	Marsupial Mammals, Creo-donta and Condylarthra, ances-tral forms of Carnivores and Ungulates, etc.
	Miocene 1,000 ft.	Not known in Britain. Touraine sands and marls, etc.	Mastodon, and early Ele-phants, Rhinoceros, Hippopota-mus, etc.
	Pliocene 200 ft.	Norwich Crag, Santa Cruz, South American	Equus, Hipparion. In South America, Edentates of various kinds. Possible appearance of man
	Pleistocene 200 ft.	Interglacial gra-vels, moraines, etc.	Palaeolithic man appears and passes by gradual transition to modern man in post-glacial times

INDEX

For EU product safety concerns, contact us at Calle de José Abascal, 56–1°,
28003 Madrid, Spain or eugpsr@cambridge.org.

www.ingramcontent.com/pod-product-compliance
Ingram Content Group UK Ltd.
Pitfield, Milton Keynes, MK11 3LW, UK
UKHW010851090126
466816UK00011B/147